David Bolius

Paleo climate reconstructions based on ice cores

David Bolius

Paleo climate reconstructions based on ice cores

Results from the Andes and the Alps

Südwestdeutscher Verlag für Hochschulschriften

Impressum/Imprint (nur für Deutschland/ only for Germany)
Bibliografische Information der Deutschen Nationalbibliothek: Die Deutsche Nationalbibliothek verzeichnet diese Publikation in der Deutschen Nationalbibliografie; detaillierte bibliografische Daten sind im Internet über http://dnb.d-nb.de abrufbar.

Alle in diesem Buch genannten Marken und Produktnamen unterliegen warenzeichen-, marken- oder patentrechtlichem Schutz bzw. sind Warenzeichen oder eingetragene Warenzeichen der jeweiligen Inhaber. Die Wiedergabe von Marken, Produktnamen, Gebrauchsnamen, Handelsnamen, Warenbezeichnungen u.s.w. in diesem Werk berechtigt auch ohne besondere Kennzeichnung nicht zu der Annahme, dass solche Namen im Sinne der Warenzeichen- und Markenschutzgesetzgebung als frei zu betrachten wären und daher von jedermann benutzt werden dürften.

Verlag: Südwestdeutscher Verlag für Hochschulschriften Aktiengesellschaft & Co. KG
Dudweiler Landstr. 99, 66123 Saarbrücken, Deutschland
Telefon +49 681 37 20 271-1, Telefax +49 681 37 20 271-0
Email: info@svh-verlag.de
Zugl.: Berne, University of Berne, Dissertation, 2006

Herstellung in Deutschland:
Schaltungsdienst Lange o.H.G., Berlin
Books on Demand GmbH, Norderstedt
Reha GmbH, Saarbrücken
Amazon Distribution GmbH, Leipzig
ISBN: 978-3-8381-1793-5

Imprint (only for USA, GB)
Bibliographic information published by the Deutsche Nationalbibliothek: The Deutsche Nationalbibliothek lists this publication in the Deutsche Nationalbibliografie; detailed bibliographic data are available in the Internet at http://dnb.d-nb.de.

Any brand names and product names mentioned in this book are subject to trademark, brand or patent protection and are trademarks or registered trademarks of their respective holders. The use of brand names, product names, common names, trade names, product descriptions etc. even without a particular marking in this works is in no way to be construed to mean that such names may be regarded as unrestricted in respect of trademark and brand protection legislation and could thus be used by anyone.

Publisher: Südwestdeutscher Verlag für Hochschulschriften Aktiengesellschaft & Co. KG
Dudweiler Landstr. 99, 66123 Saarbrücken, Germany
Phone +49 681 37 20 271-1, Fax +49 681 37 20 271-0
Email: info@svh-verlag.de

Printed in the U.S.A.
Printed in the U.K. by (see last page)
ISBN: 978-3-8381-1793-5

Copyright © 2010 by the author and Südwestdeutscher Verlag für Hochschulschriften Aktiengesellschaft & Co. KG and licensors
All rights reserved. Saarbrücken 2010

Preface

Climate change has always been a topic of great interest to myself, as it includes scientific questions as well as historical, social and political topics. During my undergraduate (master) studies in chemical engineering and analytical chemistry, however, I had barely been introduced to climatic or ecological subjects. But I had a good overview of instrumental analytic techniques and statistical methods of data interpretation. I heard of the possibilities of reconstructing climate of the past (especially temperature variations) and was fascinated by the potential of extending our climatic knowledge prior to the time of instrumental records (which roughly cover the last 150 years).

Among the best archives that have been used for this purpose are ice cores. Research on ice cores has many intriguing aspects: Exciting field work (offering the possibility of doing part of the work in an alpine environment), difficult sample preparations and interpretation of the results which is often tricky and involves highly sophisticated statistical tools. Overall, this research field seemed to offer me what I had been looking for. This work summarizes what I achieved in more than three years of investigation.

Vorwort

Das Thema Klimawandel hat mich schon immer interessiert, besonders, weil es nicht nur ein aktuelles Forschungsgebiet ist, sondern auch gesellschaftliche und politische Fragestellungen umfasst. Während meines Studiums der Technischen (Analytischen) Chemie standen allerdings ökologische Fragestellungen oder die Klimaproblematik nicht auf dem Stundenplan. Immerhin hatte ich mir ein solides Hintergrundwissen in analytischen Techniken und statistischen Methoden aneignen können. Als ich erstmals von den Methoden der Klimarekonstruktion (im Besonderen der Temperaturrekonstruktion) hörte, war ich von den Möglichkeiten, unser Wissen über den Zeitraum, für den direkte Messdaten zur Verfügung stehen (in etwa die letzten 150 Jahre), zu erweitern, fasziniert.

Eines der besten Klimaarchive, die man für diesen Zweck verwenden kann, sind Eisbohrkerne. Die Forschung, die sich damit auseinandersetzt, umfasst etliche interessante Teilbereiche: Spannende Feldarbeit, die einem die Möglichkeit bietet, neben der Büroarbeit auch in alpiner Umgebung tätig zu sein, komplexe Probenaufarbeitung und Interpretation der Ergebnisse, was oft nicht so einfach ist und den Einsatz komplexer Statistik erforderlich macht. Alles in allem schien mir dieses Feld genau jene Herausforderung zu bieten, die ich gesucht hatte. Mit der vorliegenden Arbeit stelle ich nun vor, was in mehr als drei Jahren Forschungstätigkeit herausgekommen ist.

Summary

Mountain glaciers outside the polar regions have proved to contain valuable archives recording regional climate fluctuations and atmospheric pollution history. Ice from such glaciers is well suited for the reconstruction of the concentration of species with a short residence time in the atmosphere. These species are not well recorded in polar ice sheets (Antarctica, Greenland), that are relatively far away from the source regions. From some geographic regions, however, such long term records are not yet available, although suitable glaciers principally exist there. This thesis presents results from three such glaciers: Two glaciers from South America where ice cores have been drilled and analyzed chemically for the first time, and one glacier from the European Alps.

The two glaciers in the sub-tropical Andes (33°S, Chile, and 32°S, Argentina) were investigated for their suitability as climate archives. They are located in an area of particular climatic interest as the region is strongly influenced by the El Niño climate phenomenon, which is not yet fully understood. A glacier from this region might provide a long-term archive of past variations. A reconnaissance study was conducted in 2003. Shallow firn cores were drilled and radar measurements were performed at both sites providing an estimation of the respective glacier thicknesses. After glacio-chemical analysis, the glacier in Chile was discovered to be strongly influenced by surface melting and water percolation although the drilling site was at an altitude of more than 5,300 m. Thus, this glacier was unsuitable for paleo (past) climate reconstructions.

In Argentina, a shallow firn core (13 m) was recovered from Glaciar La Ollada. Results of the analyses demonstrated the general suitability of the glacier and that melt water percolation was insignificant. The accumulation rate was 0.45 m water equivalent per year and the average $\delta^{18}O$ was -21.4‰. Snow was accumulated in summer and in winter. Due to the limited time span covered by the core (1986–2002), the existing data did not allow for reconstructions prior to available instrumental records. This project is being continued on a new deep ice core (104 m) recovered in 2005.

The second part of the thesis is devoted to the glacio-chemical analysis of an ice core that was drilled in September 2003 on Colle Gnifetti, European Alps, Switzerland, at an altitude of 4450 m (45°56′N, 7°53′E). This glacier is one of the few, if not the only site in the Alps, where the archive extends far into the past. Previous studies estimated the age of the ice at bedrock to be at least 2,000 years old. Despite the long history of ice core drilling on Colle Gnifetti (the first core was drilled in 1976), some questions have remained unanswered. A stable isotope record ($\delta^{18}O$, δD), potentially recording summer temperature variations over the last millennia, has never been published in international scientific journals. Such a record would represent a new proxy for the reconstruction of European climate variability and it could also be used for comparison with other important reconstructions based on tree ring studies.

For this study, the younger part of the 2003 core was dated by a combination of several methods: Annual layer counting, nuclear dating (^{210}Pb) and the use of reference horizons (nuclear weapon testing, Saharan dust deposition). By the application of an ice flow model based on the radiocarbon dates and other reference horizons, a continuous chronology was obtained. The theoretical basis of such flow models is discussed.

For the old part of the core, a new dating method (developed by Theo Jenk), which is based on radiocarbon measurements in carbonaceous microparticles, was applied on this core. This showed the presence of ice older than several millennia. The lowermost core section turned out to be more than 11,700 years old, suggesting the presence of ice from the last glacial period (Würm glaciation). To date, Colle Gnifetti is the glacier containing the oldest ice found in the Alps.

The accumulation rate on Colle Gnifetti was known to be comparatively low. For the ice core presented here, it was found to be between 0.36 and 0.40 m water equivalent. Borehole measurements showed temperatures ranging from -12 to -14°C.

Close to bedrock an intense yellow dust layer was observed. The layer was characterized chemically and mineralogically. Among components common for the region (e.g. quartz and feldspar) it also contained 7% of gypsum which is not found in this mountain range. This indicates that the dust is of aeolian (wind-blown) origin. It is speculated that it originates from the end of the last glacial.

Major ion records showed much higher concentration for nitrate, sulfate and ammonium during the industrial period when compared to the preindustrial period. This was attributable to increasing emissions of sulfur dioxide, nitrous oxides and ammonia. Other species such as chloride showed no trend.

A high-resolution record of stable isotopes (δ^{18}O) for the time period 550–2003 A.D. is presented and its validity for temperature reconstructions is discussed. Eventual influences of shifts in the precipitation pattern on δ^{18}O are evaluated. The highest values in the δ^{18}O record were observed during the Medieval Warm Period around 1020 A.D. The 20th century shows a strong increase in δ^{18}O of 2.1 ± 0.2‰.

Contents

1 **Introduction**   11
    1.1 Motivation for climate research 11
    1.2 Natural variability of the earth's climate 13
    1.3 Anthropogenic influence on the atmosphere and on climate . . 16
    1.4 The climate phenomenon El Niño 21
    1.5 The use of glaciers as climate archive 23
    1.6 In search for glaciers recording El Niño 27
    1.7 Contribution of this study 30

2 **Methods**   32
    2.1 Field campaigns 32
       2.1.1 Deep drilling on Fiescherhorn, December 2002 32
       2.1.2 The reconnaissance expedition to South America, January 2003 33
       2.1.3 Mercedario field campaign 2004 36
       2.1.4 Consequences of the field campaign 2004 failure 37
    2.2 Ice cutting 38
    2.3 Ion chromatography 40
       2.3.1 Blanks 43
    2.4 Stable isotope mass spectrometry 43
    2.5 Tritium 44
    2.6 Nuclear dating with lead-210 45
    2.7 Radiocarbon dating 45

3 **A first shallow firn core record from Glaciar La Ollada on Cerro Mercedario in the Central Argentinean Andes**   46
    3.1 Introduction 47
    3.2 Methods 49
       3.2.1 Drilling campaigns 49
       3.2.2 Radar survey 50
       3.2.3 Chemical analysis 51

    3.3 Results 51
    3.4 Discussion 57
           3.4.1 Seasonality of precipitation and sublimation 57
           3.4.2 Preliminary dating and net accumulation 62
           3.4.3 Influence of melt water formation and percolation . . . 64
    3.5 Conclusion 65

4 Millennial variability of European climate inferred from an Alpine ice core   71

    4.1 Introduction 71
           4.1.1 The history of ice core drilling on Colle Gnifetti 73
    4.2 Methods 75
           4.2.1 Drilling campaigns 75
           4.2.2 Chemical analysis 77
    4.3 Results and Discussion 80
           4.3.1 Bore hole temperature 80
           4.3.2 Density profile 80
           4.3.3 Dating by annual layer counting, tritium, dust layer stratigraphy, lead-210 (^{210}Pb) and radiocarbon 82
           4.3.4 Dating by flow modelling using the stratigraphic horizons and the radiocarbon ages 90
           4.3.5 Stable Isotopes 96
           4.3.6 Calibration of stable isotopes for the use as paleo thermometer 103
           4.3.7 Influence of shifts in seasonality of accumulation on δ^{18}O 109
           4.3.8 Stable isotopes as temperature proxy 111
           4.3.9 Trends over the last 1,500 years 113
           4.3.10 Comparison with temperature observations and other millennial records 115
           4.3.11 Major ion records 120
           4.3.12 An intense yellow dust layer near bedrock 123
    4.4 Outlook 127

5 Acknowledgements   128

6 Annex   131
    6.1 Bibliography - Literature 131

List of Figures

1.1 Instrumental record of global temperature from 1856 to 2004 . 12
1.2 Variations of CO_2 concentrations over the last 420,000 years . 17
1.3 Records of past changes in atmospheric composition over the last millennium 18
1.4 Variations of the EarthŠs surface temperature: years 1000 to 2100 A.D. 20
1.5 Normal conditions in the Pacific 22
1.6 El Niño conditions in the Pacific 23
1.7 Variations of δD in Antarctic precipitation reflecting temperature oscillations over the last 420,000 years 25
1.8 Concentration of lead (Pb) in a European ice core over the period 1650–1995 A.D. 28

2.1 Drilling on Fiescherhorn glacier in December 2002 33
2.2 Cerro del Plomo from camp II 34
2.3 Shallow core drilling on Glaciar Esmeralda, Cerro Plomo . . . 35
2.4 Mercedario, one of the highest summits of the Andes 36
2.5 Road buried by stones and mud from a landslide after a heavy thunderstorm the day before 37
2.6 Scheme for ice cutting applied for the Colle Gnifetti ice core . 39

3.1 Map of southern South-America 47
3.2 Map of Glaciar Esmeralda on Cerro del Plomo, Chile 52
3.3 Radar profile recorded on Cerro del Plomo, Chile, December 2000 53
3.4 Firn cores from Cerro del Plomo and Mercedario: Density increase with depth 54
3.5 Chemical and isotopic record from the firn core recovered at Cerro del Plomo 55
3.6 Map of the glacier *La Ollada* on Mercedario 56
3.7 Radar profile from Glaciar La Ollada, Mercedario 57

3.8   Chemical and isotopic record from the firn core recovered on Mercedario 58
3.9   Mercedario SO_4^{2-} concentration, $\delta^{18}O$ record and suggested annual layers 59
3.10 Cl^- records of Cerro Mercedario and Cerro Tapado 61
3.11 Seasonal cycle of precipitation at Santiago de Chile (33°S), Pisco Elqui (30°S) and San Juan, Argentina (32°S) 62
3.12 $\delta^{18}O$ records of Cerro Mercedario and Cerro Tapado 63

4.1   Location of the ice core drilling site *Colle Gnifetti* and other ice coring sites in the Alps 73
4.2   Map of Colle Gnifetti showing the estimated layer thickness of ice older than 500 years and the position of all the drilling sites there 76
4.3   The glacier saddle Colle Gnifetti as seen from Liskamm 77
4.4   Current state of the analysis on the Colle Gnifetti ice core . . 79
4.5   Borehole temperatures for the Colle Gnifetti 2003 ice core . . 81
4.6   Profile of density increase of firn and ice shown for the 2003 ice core from Colle Gnifetti 82
4.7   Annual Layer counting based on seasonal variations of $\delta^{18}O$ and ammonium concentration 83
4.8   Tritium concentration in the 2003 core and in the Blue Core . 84
4.9   A yellow layer corresponding to Saharan dust deposition in 1977 85
4.10 Matching of Saharan dust events between the well-dated Blue Core and the 2003 core 87
4.11 Sulfate/calcium ratio for the 2003 core, preindustrial section . 89
4.12 List of reference horizons obtained by different methods 92
4.13 Reference horizons and estimation of the age by a 2-parameter flow model 93
4.14 Reference horizons and estimation of the age by a flow model, suggested by Dansgaard et al. 95
4.15 Age - depth relation using a linear time axis for the age range 0–4,000 years 95
4.16 $\delta^{18}O$ record over the entire Colle Gnifetti ice core 97
4.17 Colle Gnifetti $\delta^{18}O$ record for the time period 550–2003 A.D. . 98
4.18 Colle Gnifetti $\delta^{18}O$ record of the last 10,000 years 99
4.19 Difference between pure shear and simple shear 99
4.20 Colle Gnifetti record of deuterium excess and $\delta^{18}O$ 101
4.21 Colle Gnifetti deuterium excess vs $\delta^{18}O$ for the time period 1976–2003 A.D. 101
4.22 Histogram for deuterium excess observed on Colle Gnifetti . . 102

4.23 δD vs. δ^{18}O for the Colle Gnifetti, 1976–2003 102
4.24 Linear relationship between δ^{18}O and air temperature for individual summer months June, July and August at five stations in Switzerland for the time period 1970–2004 A.D. 105
4.25 Colle Gnifetti δ^{18}O yearly averaged vs. estimated on-site summer temperature 107
4.26 Comparison of Colle Gnifetti δ^{18}O, Casty et al. temperatures and CRU temperature anomalies 1976–2000 A.D. 107
4.27 Comparison of trends in δ^{18}O and estimated temperature on Colle Gnifetti 108
4.28 δ^{18}O variability on Grimsel station, 1950 m a.s.l. 109
4.29 Simulated δ^{18}O variability for one year of accumulation 110
4.30 Simulated precipitation pattern using an annual mean of -17.6‰ and an amplitude of 6‰ 112
4.31 Colle Gnifetti δ^{18}O record 1900–2003 A.D. 113
4.32 Comparison of the effect of different smoothing algorithms on the resulting signal 114
4.33 Colle Gnifetti δ^{18}O record 550–2003 A.D. 116
4.34 Monthly surface temperature anomaly 5–10°E, 40–50°N and Colle Gnifetti δ^{18}O record 117
4.35 Comparison of Colle Gnifetti δ^{18}O and Casty et al. temperatures for 45.75°N and 7.75°E 118
4.36 Comparison of different temperature reconstructions of the last Millennium and Colle Gnifetti δ^{18}O 119
4.37 Major ions and δ^{18}O record over the entire core length 121
4.38 Position of the intense yellow dust layer near bedrock 123
4.39 Pictures of the two lowermost core segments 124 and 125 . . . 123
4.40 Calcium and sulfate concentration and δ^{18}O record around the yellow dust layer 124
4.41 δ^{18}O, calcium and sulfate concentration in the lowermost section of the Red Core 124

Unless specified otherwise, all pictures were taken by the author.

List of Tables

2.1 Components of the Dionex anion chromatograph 41
2.2 Quantitation limits for anion chromatography in μg/kg 41
2.3 Components of the Sykam cation chromatograph 42
2.4 Quantitation limits for cation chromatography in μg/kg 42
2.5 Typical concentrations found in blank ice 43

3.1 Mean concentrations of major ions and mean δ^{18}O of the shallow cores from Cerro del Plomo and Cerro Mercedario compared to values at Cerro Tapado (time period 1986–1999) . . . 54

4.1 Matching of Ca^{2+} peaks observed in the 2003 core to Ca^{2+} peaks in the well-dated Blue Core 86
4.2 ^{210}Pb activity of two core sections located below visible dust horizons at 25.7 and 39.2 m weq., respectively 88
4.3 ^{210}Pb activity at specific horizons in the 1982 Blue Core 88
4.4 Calibrated radiocarbon ages for the samples from Colle Gnifetti 90
4.5 Reference horizons used for dating 91
4.6 Dating error at different depths (m weq.) of the core 94
4.7 Deuterium excess at Swiss stations 104
4.8 Comparison of median ion concentration (μeq./L) between the 2003 core and the Blue Core 122

Chapter 1

Introduction

This chapter provides a basic introduction to the main goals and ideas behind this study in the context of the wider field of climate research. Therefore, the motivation for climate research is explained and a brief summary is given about the evolution of climate before human influence. It will be made especially clear why understanding the natural variability of climate is so important and how this study intends to contribute to this knowledge. Then, the influence of human activity on climate is addressed and the possible consequences for the future. Finally, the specific topics relevant to my research are introduced to facilitate the interpretation of the presented results.

1.1 Motivation for climate research

Because anthropogenically induced climate variability is always superimposed on (or "mixed with") natural variability, disentangling the two signals is one of the main tasks in climate research. The better each of these contributions can be understood and quantified, the more accurate prediction for future climate change will be. Based on such predictions, it will be possible to react to changing climate by adaptation (coping with new boundary conditions) and mitigation (trying to reduce the magnitude of change). Comparing e.g. the impact of different greenhouse gas emission scenarios on future climate change might provide useful evidence to convince policymakers to take certain measures.

There is abundant evidence for global warming in the 20th century. According to instrumental data global mean temperature rose by 0.6 °C (see Fig. 1.1)[1] during that period. However, it is still not possible to exactly

---
[1] The data are available (Dec., 2005) from
http://www.cru.uea.ac.uk/cru/info/warming/

disentangle and quantify the fraction of natural and anthropogenic forcing, respectively. Moreover, good instrumental data is only available for the last ~150 years such that knowledge about earlier climatic variability relies on the use of so called *archives* (see Section 1.5). Climate change skeptics, who

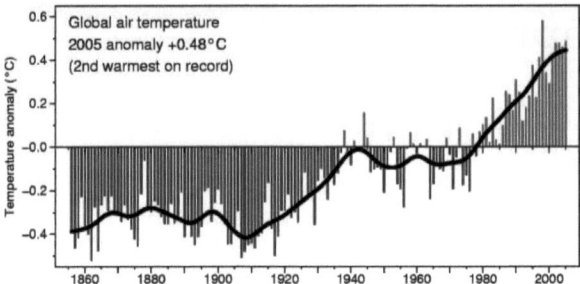

Figure 1.1: Instrumental record of global temperature from 1856 to 2004. The graph is adapted from Jones et al. (2003).

nowadays are a minority in the scientific community, use this weak point as an argument for denying any anthropogenic forcing and ascribe the observed warming to solely natural variability or they doubt the reliability of the data. Today, the three main topics in climate research can be summarized by:

- Understanding the climatic response to anthropogenic forcing

- Understanding climate dynamics and natural climatic variability of the past

- Predictions of *future climate*

Obviously, these topics are intimately linked to each other and all require interdisciplinary research. Reconstruction of paleo climate therefore contributes to the whole understanding of how the climate system works and thus enhances the performance of models predicting future climate of the 21st century. The essential link between climate reconstruction and its prediction was well summarized by a famous British statesman, who probably did not have climatology in mind when he said:

> The farther backward you can look, the farther forward you are likely to see

Sir *Winston* Leonard Spencer *Churchill* (1874–1965)

In other words, in order to understand future climate variability and its response to anthropogenic forcing one has to maximize the understanding of past oscillation, a topic which is discussed in more detail in the next Section (1.2). Moreover, only someone who is aware of the entire range of natural fluctuations and their consequences for life on earth and also of the rate at which changes occur, will be able to assess the impact of future change.

The purpose of this study is to provide a small contribution to the growing knowledge about past climatic variability by bringing up results of several climate archives: two glaciers in South America, where cores were drilled for the first time, and a glacier in the Alps. Thus, new climatic information is made accessible for time periods instrumental data is not available for. But, before going into detail, a short summary about climatic history is presented.

1.2 Natural variability of the earth's climate

The climate of the earth has varied naturally on all time scales. These variations include trends over millions or thousands of years to short-term decadal variability. Over the course of the last billion (10^9) years climatic conditions have changed between a planet that was completely frozen (Snowball Earth, ~600 Million years ago, Hoffman et al. (1998)) and extremely warm greenhouse conditions during Cretaceous (the era of the Dinosaurs), 100 Million years ago. However, the existence of the *Snowball Earth*, including glaciated tropics, is not entirely beyond doubt for parts of the scientific community.

During the Cretaceous, ice did not exist anywhere on earth, even Antarctica was an ice-free, forested continent although it was close to its current position centered on the South Pole. The dinosaurs even lived close to the polar circle under subtropical conditions (e.g. Chinsamy et al., 1998; Fiorillo and Parrish, 2004). Over the last 50 million years the earth's climate has undergone a long term cooling trend (e.g. Zachos et al., 2001; Ruddiman, 2001; Schönwiese, 1992). Between 34 and 40 million years ago, a step towards an "ice house" earth was made when a permanent ice sheet built up in Antarctica (Zachos et al., 2001) that accounted for half of the mass of today's ice there (Zachos et al., 1993). The Northern hemisphere, however, would still remain ice-free for another 50 million years. The exact reason for the persisting cooling trend is unknown but a plausible theory strongly propagated by William F. Ruddiman suggests that the main driving factor was the collision of two continents, India and Asia starting around 55 million years ago (Ruddiman, 1997) and initiating the uplift of the Himalaya mountains. This process is still going on today.

The existence of high-elevation terrain is linked to global climate by the process of chemical weathering (erosion) of silica rock. This reaction is a sink of atmospheric CO_2, taking place everywhere on earth. Out of the thousands of different types of silicate minerals on earth, the weathering process is shown for a simple structure, wollastonite (simplified[2], Ruddiman (2001, p. 94)):

$$CaSiO_3 + H_2CO_3 \longrightarrow CaCO_3 + H_2O + SiO_2$$

Within the carbon cycle, this process accounts for most of the carbon that is transferred from the atmospheric reservoir into carbonate rock (0.15 gigatons per year). It is interesting to note that the chemical weathering rate on earth depends on the type of terrain and it is elevated by a factor of ~50 in high mountains when compared to the rate in Basins (Ruddiman, 2001, p. 167). Thus, the uplift of the Tibetan Plateau and the Himalaya is thought to have increased chemical weathering on a global scale and to have caused a consequent drop in atmospheric concentration of the important greenhouse gas CO_2. This long-term decrease is probably responsible for the cooling trend over the last 50 million years. Around 2.7 million years ago, it had become cold enough to allow for perennial ice to persist in the Northern Hemisphere and the huge ice cap on Greenland came into existence (Haug et al., 2005). This boundary marked the start of a new geologic era, the so called *Quaternary*.

By that time, humans started to populate the planet, although for a long time they mainly lived in Africa. The earth's climate had become highly variable and relatively cold. On time scales of several tens of thousands of years it had started to oscillate between cold glacials and relatively warm interglacials. The ultimate causes and mechanisms triggering these oscillations are not yet fully understood, however it was postulated by Milutin Milankovitch (1941), a Serbian astronomer[3], that variations in the earth's orbit around the sun and subsequent fluctuations in the heat received over specific

---

[2] The reaction involves several steps:

1. $H_2O + CO_2 \longrightarrow H_2CO_3$ Formation of carbonic acid
2. $CaSiO_3 + 2H_2CO_3 \longrightarrow Ca^{2+} + 2HCO_3^- + H_2SiO_3$ Dissolution of silica rock
3. $H_2SiO_3 \longrightarrow H_2O + SiO_2$ Silica formation
4. $Ca^{2+} + 2HCO_3^- \longrightarrow CaCO_3 + H_2O + CO_2 \uparrow$ Carbonate precipitation in the ocean

CO_2 is released again in the last step but only 50% of what was initially taken up in step 2. Thus, the process is still a net sink for this greenhouse gas.

[3] The life of M. Milankovitch and the discovery of the ice ages is very well narrated in "Ice Ages - Solving the Mystery" (Imbrie and Imbrie, 1979). See Section 6.1, p. 131, Recommended Reading

areas of the globe cause the transitions between glacials and interglacials. We currently live in an interglacial, the so called *Holocene*. During an interglacial the distribution of ice on the continents is comparable to today with large ice caps only existing near the poles in Antarctica and Greenland, respectively. Contrastingly, during a glacial, massive ice sheets cover North America and Scandinavia, global mean temperatures are by $\sim 4°C$ lower than today (CLIMAP, 1981) and in response to that the vegetation cover looks completely different.

Since the end of the last glacial (Würm glaciation) 12,000 years ago, humans have lived in an interglacial (the Holocene). Compared to the highly variable climate that is encountered during glacials, the Holocene has been a period of relative stability. Most intriguingly, the transition from the stone age to the modern industrial age has entirely taken place during this brief period of relatively stable climate which made reliable agricultural food production available. People started to settle down and give up their former lives as hunters and gatherers. Evidence for a first permanent settlement (the first "city") in Jericho dates back to $\sim 11,000$[4] years ago and the Egyptian and Babylonian civilizations showed up about 5,000 and 4,000 years ago, respectively.

Although climatic variations during the Holocene have been small compared to those occurring during glacials, they have not been insignificant, e.g. great parts of the Saharan desert, which are now hyper-arid, were covered by vegetation in the early Holocene. Moreover, the level of north-African lakes stood much higher then. Climatic variations during the Holocene have also strongly affected human societies. By comparing historic evidence to knowledge about past climatic variations it was found that the collapse of ancient civilizations could sometimes be directly related to climatic variability, especially when a transition towards drier conditions took place. This was observed e.g. for the decline of the Akkadian Empire (Mesopotamia, Cullen et al. (2000)) around 2170±150 BC and the demise of the Maya empire (northern Central America) after the occurrence of multi-year droughts centered at approximately 810, 860, and 910 A.D. and leading to social stress (Haug et al., 2003).

---

[4] http://en.wikipedia.org/wiki/Jericho (January, 2006)

1.3 Anthropogenic influence on the atmosphere and on climate

Whereas climatic variations have affected humans through much of their history it is estimated that humans have had influence on climate for only a relatively short period of time. It was the beginning of the industrialized era (between 1750 and 1850) that marked the start. By the combustion of fossil fuels (coal, oil and gas), humans started to alter the composition of the atmosphere, especially greenhouse gas concentrations, and the rate of change has been increasing since then. Maybe the alteration of atmospheric greenhouse gas concentrations started much earlier. Methane concentration began to rise 5,000 years ago and Ruddiman and Thomson (2001) suggested that this might have been due to rice cultivation by early human populations. However, the increase is small when compared to the rise since the industrial revolution.

The greenhouse effect is a process that scatters back some of the infrared (heat) radiation that the earth would otherwise directly emit into space. Therefore, it keeps the planet warmer than it would be if the heat was lost. Today, the global mean temperature is ~15°C whereas it would be only ~-18°C without this extra warmth (e.g. Schönwiese, 1992, p. 136). The main natural greenhouse gases are water vapor (H_2O), carbon dioxide (CO_2), ozone (O_3) and methane (CH_4). Hence, humans have not introduced the greenhouse effect on earth, but they are reinforcing it by increasing atmospheric concentrations of the respective gases.

Over the last 420,000 years, pre-industrial concentration of one of the most important greenhouse gases, CO_2, had oscillated between 280 ppm (parts per million) during warm interglacials like the Holocene and 180 ppm during cold glacials (see Fig. 1.2). Since the industrial revolution it has increased to 380 ppm (2004)[5]. The more recently recovered antarctic EPICA[6] Dome C ice core has extended the greenhouse gas record now covering 650,000 years (Siegenthaler et al., 2005; Spahni et al., 2005). However this has not changed the general picture of the glacial-interglacial changes even though the amplitude of change was reduced during the time period 420,000–650,000 before present.

CO_2 is not the only trace gas that has had rising concentrations since the

---

[5] Atmospheric CO_2 concentrations have been recorded at the Mauna Loa observatory, Hawaii, since 1958. The data is available (January, 2006) at http://cdiac.esd.ornl.gov/ftp/trends/co2/maunaloa.co2

[6] EPICA: European Project for Ice Coring in Antarctica, Augustin et al. (2004); Wolff et al. (2006)

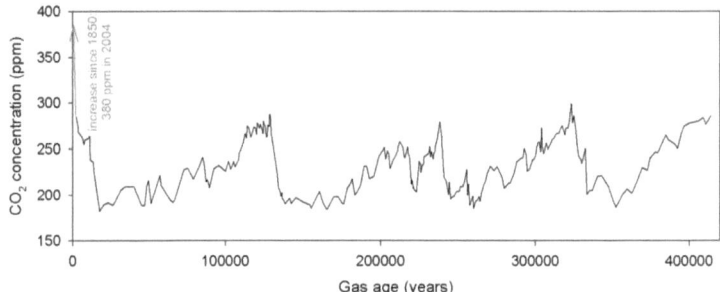

Figure 1.2: Variations of CO_2 concentrations over the last 420,000 years as inferred from the Antarctic *Vostok* ice core and anthropogenic increase since the industrial revolution (~1850). The composition of the atmosphere is directly archived in air bubbles enclosed in the ice. The data is adapted from Petit et al. (1999).

industrial revolution. The fraction of methane in the atmosphere, yet another important greenhouse gas, has doubled since 1750 (Fig. 1.3). Its main sources are related to rice production and cattle-breeding. Another pollutant, that is a byproduct of fossil fuel burning, is SO_2, which was also emitted at an increasing rate. This gas does not play an important role as a greenhouse gas but in the atmosphere it is oxidized and converted to sulphuric acid, H_2SO_4, which was the most important compound involved in the acidification of rain observed in the 1970s and 1980s. Sulphate (SO_4^{2-}) concentrations in (alpine) ice cores are a good tracer for past emissions. They peaked in the 70s have decreased since then (Fig. 1.3), mainly due to the desulphurization of fuel.

One of the most intriguing problems is the quantification of the magnitude and the impact of future climate on humanity which makes prediction a very important issue. Since the 1970s climate scientists have increased their ability to model global climate on computers. This has to be considered a real milestone as it is the only way humans are able to quantitatively make predictions about future climatic variations. The only other way of prediction would be projection of past climate into the future. This will not work as there is *no* climatic analogue in the past for the boundary conditions and the rapidity of change humanity is going to face in the 21st century!

According to possible emission scenarios of the IPCC[7] (IPCC - Synthesis

---

[7]IPCC - Intergovernmental Panel on Climate Change. The English version of the *Synthesis Report* is available at http://www.ipcc.ch/pub/syreng.htm (January, 2006)

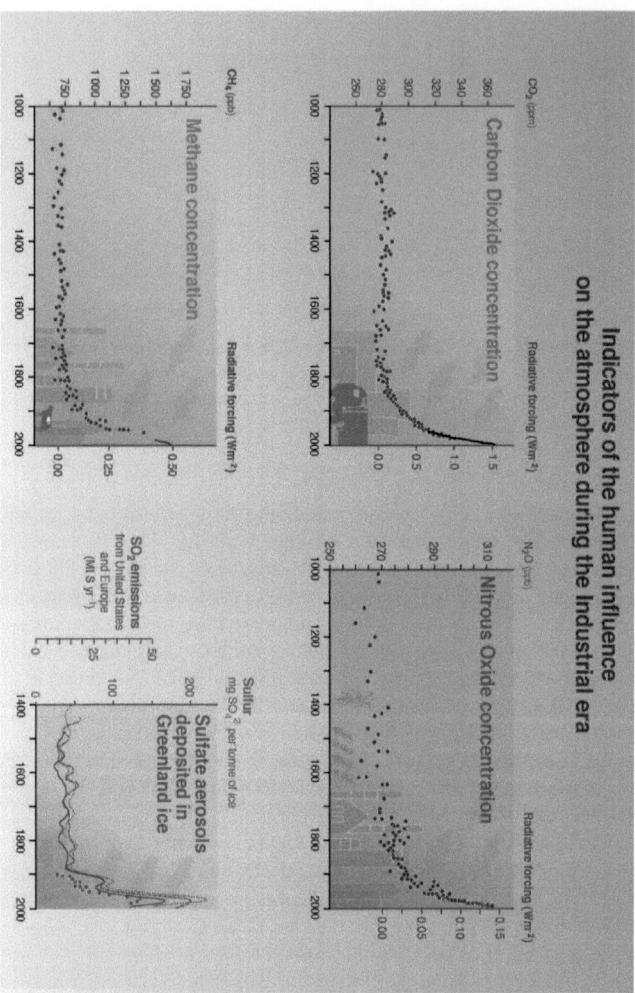

Figure 1.3: Records of past changes in atmospheric composition over the last millennium demonstrate the rapid rise in greenhouse gases and sulfate aerosols that is attributable primarily to industrial growth since 1750. From IPCC - Synthesis report (2001, p. 47).

report, 2001, p. 60) the concentration of CO_2 will continue to rise strongly and reach 540 to 970 ppm by the end of the 21st century. The authors warn that this will trigger a significant increase of the global mean temperature between 1.8 to 5.8 °C (IPCC - Synthesis report, 2001, p. 61). This is a huge increase when compared to the observed warming in the 20th century (0.6°C). It is also in the order of the temperature difference between full glacial and present-day conditions which was estimated to be ~4°C (CLIMAP, 1981). Figure 1.4 shows variations of the Earth's surface temperature from 1000–2100 A.D. for various scenarios based on different economic and population growth and technological progress.

The expected climate change will also be characterized by an increasing frequency in the occurrence of extreme events. Especially heat waves have a much higher probability to occur in a warmer world. Such a heat wave was experienced in Europe in 2003, with June, July and August temperatures in Switzerland being 5.1°C warmer than on average (Schär et al., 2004). This corresponded to more than a 5σ deviation. By 2071–2100, every second summer is expected to be at least as warm as summer 2003. The extreme temperatures associated to this event caused an increased mortality especially among elderly people. The excess mortality was estimated to be ~14,800 in France during August 2003 (Dhainaut et al., 2004).

The sea level is believed to rise by 9 to 88 cm (IPCC - Synthesis report, 2001, p. 64) during the 21st century. This is a direct consequence of higher air temperatures that cause the ongoing retreat of glaciers and ice sheets and the thermal expansion of sea water. The large uncertainty in this estimation is due to difficulties in estimating how much of the polar melting will be compensated by increased snowfall there. Although 9 cm does not sound like much it is important to note that "The most serious impacts are caused not only by changes in mean sea level but by changes to extreme sea levels, especially storm surges and exceptionally high waves, which are forced by meteorological conditions" (IPCC - The Scientific Basis, 2001, p. 644).

Even on a political level it is increasingly realized that climate change might become a major challenge to humanity in the 21st century. Most countries recognize greenhouse gas emissions (mainly CO_2) as the main reason for climate change. In 1997, the *Kyoto Protocol*[8], negotiated in Kyoto, Japan, was adopted as amendment to the *United Nations Framework Convention on Climate Change (UNFCCC)*. The Protocol was the first treaty to contain binding definitions on the reduction of greenhouse gases below the emission level of 1990. This target shall be achieved by the year 2012.

---

[8]http://de.wikipedia.org/wiki/Kyotoprotokoll and
http://en.wikipedia.org/wiki/Kyoto_Protocol (January, 2006)

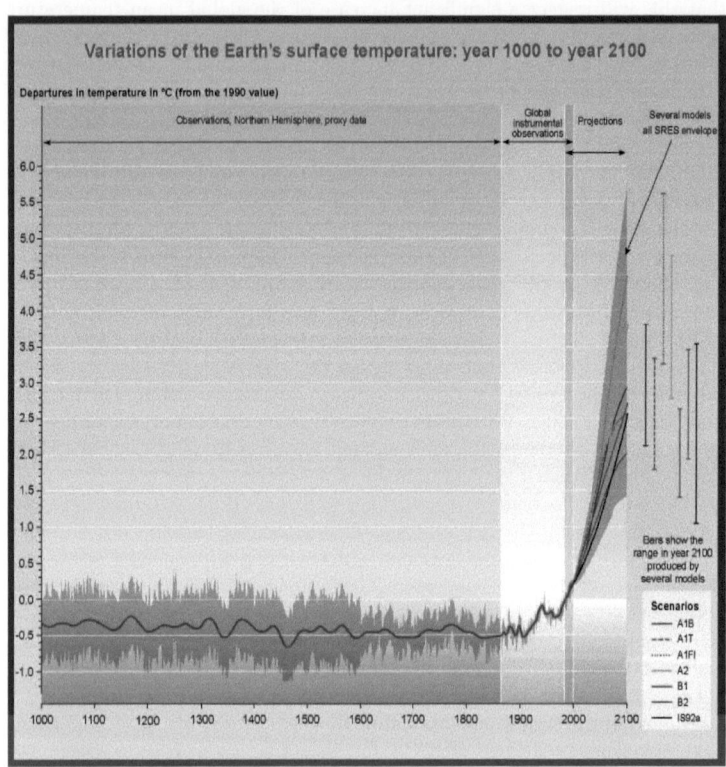

Figure 1.4: Variations of the EarthŠs surface temperature: years 1000 to 2100 A.D. For the period 1000–1860 the values are reconstructed using different climate archives (tree rings, corals, ice cores and historical data). For 1860–2000 temperature is based on observational data and the period 2000–2100 shows projections based on six different scenarios. The figure was taken from IPCC - Synthesis report (2001, p. 140).

Countries of the European Union e.g. have agreed to a cut, on average, by 8% from 1990 emission levels. The United States of America signed the Protocol under president Clinton but the current administration under president G.W. Bush has no intention to ratify it. China and India signed the treaty but as both nations were classified as "developing countries", they have no obligation to reduction. The treaty only entered into force in 2005 after it had been ratified by Russia. This fulfilled the condition that enough countries responsible for at least 55% of the 1990 worldwide CO_2 emissions had gone through the ratification process. However, contrary to the intentions of the treaty, greenhouse gases are now expected to rise by 11% for 1990–2010. This makes it look unlikely that the goals can be achieved, however, it is too early to give a final statement.

Some people even rank climate change as the top problem of the world. Among them is *Sir David A. King*, U.K. Chief Scientific Advisor, who stated (2004):

> Climate change is the most severe problem that we are facing today— more serious even than the threat of terrorism

It remains hard if not impossible to give an objective estimation of what the biggest danger to humanity will be. Although climate change is definitely a very serious issue, wars, diseases, overpopulation and limited resources should not be overlooked, either. It is therefore a question that cannot be answered in a general way.

1.4 The climate phenomenon El Niño

El Niño is a very important phenomenon strongly affecting climatic conditions in the Pacific (e.g. wind patterns, precipitation distribution, temperatures). It consists of a certain "extreme" state of the coupled ocean-atmospheric system: Under "normal" conditions a strong cyclone (low pressure) system is located over Indonesia and anticyclones (high pressure systems) are found in the eastern Pacific north and south of the equator, respectively (see Fig. 1.5). Consequently, there is a ground-level air flow from the centers of high pressure towards the Indonesian low pressure system. On their way, the air masses are deflected towards the west. These relatively stable air flows are called the *Trade Winds*, as they used to provide the sailing power for early tradesman on their way from Europe to the New World (these winds also blow in the Atlantic). The trade winds exert a force on oceanic surface waters and thus drive a strong current which has its source regions along the coasts of the Americas. These coasts are regions of upwelling deep ocean water that is cold and rich in nutrients.

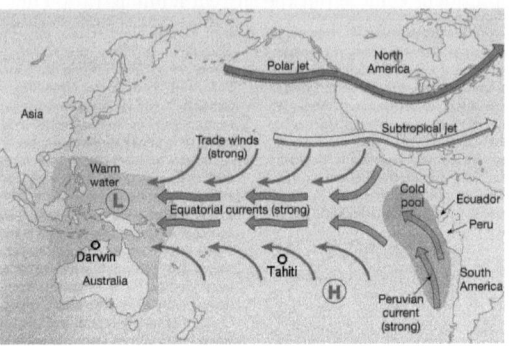

Figure 1.5: Normal meteorological conditions in the Pacific. Near the equator, strong trade winds blow from the east (thin arrows) and drive an ocean current (thick arrows). Sea surface temperature within this band increases from east to west. A low-pressure system is located over Indonesia, and high-pressure prevails over French Polynesia (Tahiti). Figure from Lutgens and Tarbuck (2001).

During El Niño conditions the boundary conditions are different (see Fig. 1.6). The pressure gradient (difference) between the subtropical anticyclones and the Indonesian low-pressure system becomes smaller. Consequently, the trade winds weaken and provide too little power to maintain the normal ocean current, which then reverses and brings warm surface water to the American coasts. There, the amount of upwelling deep water is dramatically reduced.

The deviations from normal meteorological conditions during El Niño imply several consequences:

- Because the Indonesian cyclone is less pronounced there is a general decrease in precipitation over Indonesia and Australia (low-pressure systems are generally associated with strong precipitation)

- Over equatorial South America rainfall is also reduced, in this case because weaker trade winds advect less moisture from the Atlantic Ocean

- In subtropical South America, between 28°S and 35°S, El Niño years are associated with increased precipitation (Aceituno, 1988). This is

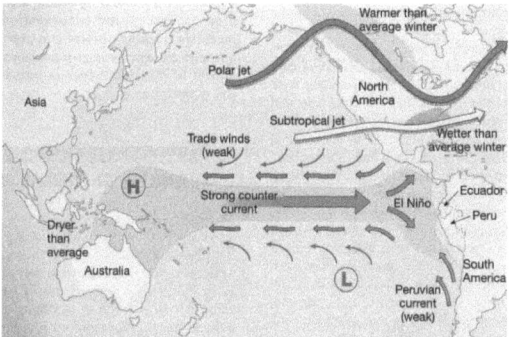

Figure 1.6: El Niño conditions in the Pacific. Trade winds are much weaker and the ocean current is inverted. Air pressure is much higher over Indonesia, causing below-average precipitation, and it is lower over French-Polynesia. Figure from Lutgens and Tarbuck (2001).

probably attributable to westerly winds reaching further north because of a less pronounced anticyclone (the Westerlies are responsible for moisture advection from the Pacific)

Two parameters are widely used for describing the conditions in the Pacific. First, the normalized surface air pressure difference between Tahiti (French Polynesia, 17°40′S, 149°30′W) and Darwin (Australia, 12°27′S, 130°50′W, see Fig. 1.5) is used and referred to as the *Southern Oscillation Index* (SOI). Under normal conditions, when the high-pressure system is strong, SOI equals zero. During an El Niño, when the anticyclone weakens, the SOI has negative values. The second parameter used is the standardized *Sea Surface Temperature* (SST) in several regions of the equatorial Pacific. Under El Niño conditions, when the supply of cold water from the American coast diminishes, sea surface temperatures are higher, thus SST is positive. Because both things happen simultaneously, sea surface temperatures and the Southern Oscillation Index are strongly (and negatively) correlated. Therefore, this phenomenon is referred to as the *El Niño-Southern Oscillation* (ENSO).

1.5 The use of glaciers as climate archive

The exploration of past climate prior to instrumental records is based on the interpretation of climate archives. Several of these archives exist and all of them contain so called *proxies* that are in some way related to climate. For example, in areas where the growth of trees is limited by low temperatures, the annual ring width is strongly related to the temperature during the growing season of that specific year. However, interpretation of these proxies is in most cases not straight forward, as they might also depend on completely different factors. Young trees grow much faster than old trees and if this is not accounted for during the interpretation, warmer climate will be assumed during the time, when the tree was young. There are many other climate archives that have played an important role:

- Historical data
- Ice sheets and glaciers
- Tree rings
- Speleothems
- Lake and ocean sediments
- Subsurface temperature profiles obtained from borehole measurements

All these archives have their individual weaknesses and strengths such that it is often helpful to combine many of them ensuring redundant information. This strategy has recently become very popular and is referred to as *multiproxy* approach (e.g. Moberg et al., 2005). Ice cores are definitely among the best existing archives because they provide many different parameters and cover a useful time range (up to 740,000 years, Augustin et al. (2004)) at high temporal resolution. Ice cores allow for the reconstruction of

- Gaseous composition of the atmosphere (including concentrations of greenhouse gases[9] like CO_2, CH_4 and N_2O)
- Ambient aerosol concentration and composition (including e.g. mineral dust)
- Past temperatures and changes in atmospheric circulation
- Changes in precipitation amounts

No other climate archive allows for the simultaneous determination of such a variety of different parameters. Most of the ice cores have been retrieved in Greenland and Antarctica with the new EPICA core and the NGRIP core

---

[9]See the data for an Antarctic ice core in Figure 1.2 on page 17

(North Greenland Ice core Project, Andersen et al. (2004)) being the most recent examples.

Ice core drilling has a long tradition in polar regions and started in 1960 with the recovery of an ice core in Camp Century, Greenland[10]. American soldiers were stationed there in a military base that had been dug into the ice and it even had its own portable nuclear power plant providing energy and heat. They were waiting for an attack by the enemy but as this did not happen, the engineers started to drill into the ice underneath (Sune O. Rasmussen, personal communication). After three attempts they could recover a core reaching bedrock at 1387.4 m. This core was analyzed among others by a Danish scientist, W. Dansgaard, who had become renowned for his work on stable isotopes in precipitation (publishing his well known paper *Stable Isotopes in precipitation* in 1964). He realized the potential of these ice cores and he was for a long time one of the leading scientists in this field.

Until recently, the most famous ice core record from Antarctica was the Vostok core drilled at a Russian research station. Figure 1.7 shows the vari-

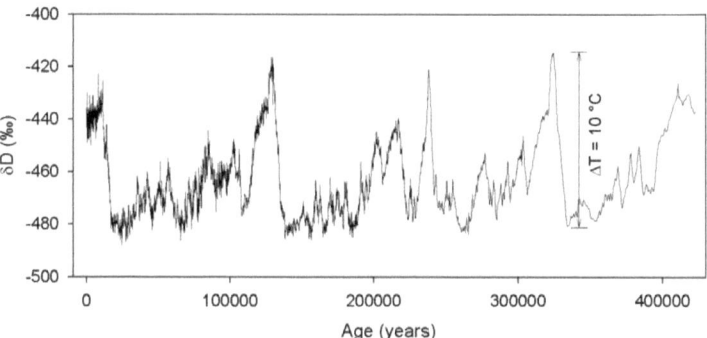

Figure 1.7: Variations of δD in Antarctic precipitation over the last 420,000 years inferred from the *Vostok* ice core. δD is an excellent proxy for past temperature variations such that lower values reflect lower temperatures. The data is adapted from Petit et al. (1999).

ation of δD[11], a very good proxy for temperature, in Antarctic precipitation

---

[10]See http://www.ncdc.noaa.gov/paleo/icecore/greenland/gisp/campcentury/campc_data.html

[11]δD is explained in more detail in Section 2.4, p. 43

over the last 420,000 years. Low δD values in Figure 1.7 indicate low temperatures. Such periods (δD < −450‰) thus correspond to long glacials and were interrupted by short interglacials roughly every 100,000 years. During the most intense phase of such glacials Antarctica was $\sim$10°C colder than during an interglacial and, at the same time, huge ice sheets covered North America and Scandinavia. The Holocene, the ongoing interglacial we live in, has lasted for 12,000 years. During the last interglacial, the Eemian, temperatures were slightly higher in Antarctica. The transition from interglacial to glacial conditions is relatively slow and takes tens of thousands of years (as does the buildup of continental ice sheets), whereas going from full glacial to interglacial conditions is abrupt implying temperature increase of $\Delta T = \sim 10°C$ in 5,000–10,000 years. As the melting of continental ice sheets was slower than the increase in temperature, they persisted into the early Holocene and their last remnants in North-America disappeared only 6,000 years ago (Ruddiman, 2001, p. 303).

Later, ice coring was extended to mountain glaciers in the mid- and low latitudes (non-polar regions). These glaciers are remarkably different to those in polar regions. They generally experience higher annual accumulation (typically 0.5–2.5 m water per year) and their thickness in the accumulation area is typically 100–300 meters which is 10% of the thickness of the ice sheets in Greenland and Antarctica. Therefore, they generally cover a much shorter time period ranging from less than 110 years as observed for an ice core from Col du Dôme, France (covering 1890–1994, Preunkert et al. (2003)) to more than a glacial cycle (100,000 years) for an ice core from the Tibetan Plateau (Thompson et al., 1997), but such old ice was only found once outside the polar regions. The higher accumulation rate on mid- and low latitude glaciers has also the advantage of allowing sub-annual resolution. The type of recorded signal is much more local compared to polar glaciers which archive climatic variations on a global or hemispheric scale. Therefore changes in regional moisture and temperature are assessable, information that cannot be gained from polar ice cores. Another strength of non-polar glaciers is that they are located much closer to sites where humans live and where the centers of industrialization are. Mountain glaciers from such regions provide records of anthropogenic air pollution especially for species with a short residence time in the atmosphere (e.g. sulphate, lead, black carbon).

Although a few studies had been published earlier, the real breakthrough for ice coring on mountain glaciers was recognized by 1985, when Lonnie G. Thompson published the results of the Quelccaya ice cap, Peru, in tropical South America. He showed the potential of such glaciers to record past El Niño events (Thompson, 1984) and he used the obtained record to explain variability in precipitation trends over the last 1,500 years (Thompson et al.,

1985). In this region, changes in the evaporation-precipitation cycle are of special importance as the tropics are a key driver for atmospheric circulation over the entire planet. The first non-polar ice core that extended back into the last glacial (covering 19,000 years) was again drilled in Peru, on Huascarán, the highest summit of the Cordillera Blanca (Thompson et al., 1995). The interpretation of the record implied sea surface temperatures in the tropical Atlantic, the moisture source for Huascarán, to be lower by 5° to 6°C during the late glacial, whereas previous studies had only claimed temperature anomalies ranging from 1° to 2°C. The discrepancy between these different estimates is still not fully resolved.

Moreover, mountain glaciers are excellent archives for anthropogenic air pollution. This was for example well documented for the history of lead (Pb) pollution which had its maximum in the 1970s when tetra-ethyl lead, a very toxic substance, was widely used as an additive for gasoline. In the 1980s a drastic decrease in atmospheric lead concentrations was observed in Europe (and North America), especially after the introduction of catalytic converters for cars because these cars could not be run with leaded gasoline. This decrease is obvious from Figure 1.8 showing lead concentration in an ice core from Colle Gnifetti. However, leaded gasoline was only recently phased out in China (around 2001). It has only been possible to measure lead in the environment directly for the most recent decades. This underlines the importance of extending our knowledge to time periods direct measurements are not available for. Glaciers are excellent archives for this purpose. Accurately documenting the impact on environment and health illustrates the benefit and efficiency of air pollution control.

1.6 In search for glaciers recording El Niño

The goal of the first part of this study was to find a glacier that would provide an archive for past El Niño events. Such an archive would help to understand this phenomenon, which represents the strongest natural fluctuation of present day climate (further details about this phenomenon were given in Section 1.4). If such a glacier could be found in a region that is influenced by the El Niño-Southern Oscillation (ENSO) it could potentially provide such an archive.

South America is an ideal place for retrieving ice cores: The Andes extend over 7,000 km from the north to the south of the continent and thus allow for the existence of glaciers at altitudes of more than 6,000 m. Together with the mountains of North America, glaciers between the two poles are available allowing for a transect study Pole-Equator-Pole, called PEP 1 (Markgraf

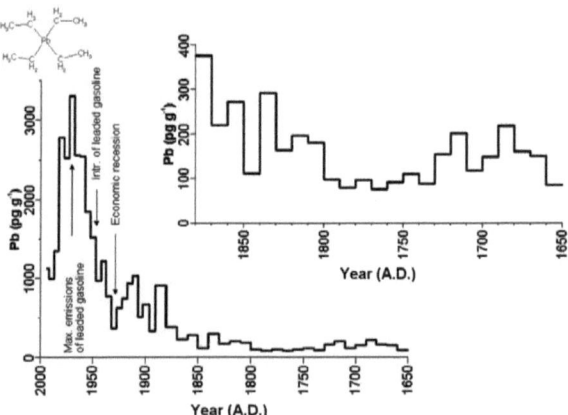

Figure 1.8: Lead (Pb) paleo concentration record from the Colle Gnifetti as 5-year (period 1890–1995 A.D.) and 10-year averages (period 1650–1890 A.D.). The period from 1650 to 1880 A.D. is additionally shown with a magnified horizontal axis, to take into account the lower lead concentrations. The molecular structure of tetra-ethyl lead is shown in the upper left corner. Adapted from Schwikowski et al. (2004).

et al., 2000).

South America's precipitation pattern is characterized by tropical summer precipitation in the north (tropical South America), where the moisture source is the tropical Atlantic, and the west wind belt with winter precipitation in the south (extra-tropical South America). Here the moisture comes from the Pacific. The intersection of the two regimes is called the "dry diagonal" (see Figure 3.1 on page 47) and there is practically no precipitation at all (Atacama desert). The boundaries of this region are expected to be most sensitive to any changes in the moisture regime. So far, ice core records that have been interpreted in terms of the El Niño-Southern Oscillation, were obtained in tropical South America on Quelccaya (Thompson, 1984) and Huascarán (Henderson et al., 1999) and the section interpreted for ENSO was very limited (1964–1983 and 1925–1993, respectively). It would therefore be desirable to retrieve an ice core from the Pacific side of sub-tropical South America, where the imprint of El Niño is potentially stronger, because ENSO is a phenomenon of the Pacific and here the moisture source is the Pacific. In addition, extension of the ENSO-related proxy record would be advantageous.

Previous studies in the context of ENSO reconstructions based on ice cores from the extra-tropics have already been carried out in the frame of the SNF project "Paleoclimate of the Central Andes" (SNF 21-50854.97/.99) lead by Prof. H. Veit, University of Berne. An ice core from Cerro Tapado (5536 m a.s.l., 30°S, Chile) was investigated by P. Ginot (Ginot, 2001) in the hope that the site would represent a good climate archive for reconstructing ENSO variations of the past. However, the core only covered the time period 1920–1999 (Ginot et al., 2005). Cerro Tapado is very close to the dry diagonal and sublimation was found to be very strong here under high insulation and low relative humidity (2–4 mm water loss per day Stichler et al. (2001)). Sublimation influenced the stable isotopes composition (δD, $\delta^{18}O$) and lead to enrichment in the surface layer but the enriched signal did not penetrate into deeper firn layers (Stichler et al., 2001). It was also shown that sublimation and dry deposition strongly influenced the chemical concentrations in that record such that conservative species were enriched at the surface. This was demonstrated by surface snow experiments (Ginot et al., 2001a). Post-depositional effects such as sublimation or surface melting hinder the interpretation of ice core records. Hence, a site where these effects are weaker is preferable.

1.7 Contribution of this study

This dissertation presents the results of the analysis of three ice cores, two shallow (short) cores from glaciers in the Andes and one deep core drilled to bedrock on a glacier in the Alps.

Shallow cores retrieved in South America

Because of the limitations of previous studies (Section 1.6), the goal of this work was to discover a new suitable glacier archive, potentially recording ENSO variations of the past. Therefore, the obvious next step was to search for another site where the record could be extended farther back in time. The glacier should be located south of Cerro Tapado (almost no glaciers exist north of Cerro Tapado where it is even drier) and therefore be less influenced by sublimation as humidity and precipitation increase further south (Figure 3.1 on page 47).

In January 2003, a field campaign was conducted in search for such a glacier (more details about the campaign are given in Section 2.1.2). The light-weight, easily portable drilling system FELICS small (Ginot et al., 2001b) was used and shallow firn cores were successfully recovered from two glaciers in Chile and Argentina, respectively. They were analyzed (Chapter 2, Methods) and the results, presented in Chapter 3, looked very promising for the Argentinean glacier (La Ollada).

Another drilling campaign, to recover of a deep core to bedrock was conducted in February 2004. Unfortunately, this drilling campaign failed due to persistent bad weather conditions. Consequently the topic of this thesis had to be extended to another subject as there was no ice to analyze.

A new deep ice core from Colle Gnifetti, drilled in 2003

The second part of this study is devoted to a deep ice core drilled in September 2003 on Colle Gnifetti (4454 m a.s.l., 45°56′S 7°53′E), Swiss Alps. Because of the low accumulation rate on this site (0.40 m, see Section 4.3.4), the glacier was regarded to be one of the few, if not the only site in the Alps, bearing old ice (500–10,000 years) which would provide a long-term climatic record. Cores had been retrieved here previously, but a record of stable isotopes ($\delta^{18}O$)[12] had never been published in international journals although an entire dissertation devoted to the interpretation of $\delta^{18}O$ exists (Keck, 2001). Stable isotopes in water represent the parameter which is most closely linked to climatic variations, because under certain conditions $\delta^{18}O$

---

[12] Information about stable isotopes in water is given in Section 2.4 on page 43

is a good proxy for past temperature variations. Therefore, one of the main goals of this project was to provide a long-term record of climatic variability based on an ice core from Central Europe.

Chapter 2
Methods

This chapter shall give an overview on how ice core records were obtained. It therefore gives a chronology of all the different steps: Drilling on high-mountain glaciers, transport of the samples (Section 2.1), ice cutting and preparation for analysis (Section 2.2), as well as analytical methods used such as ion chromatography (Section 2.3) and stable isotope mass spectrometry (Section 2.4). In total more than *600* samples were analyzed for major ions and stable isotopes related to the South America project and more than *1600* samples were analyzed from the Colle Gnifetti ice core.

2.1 Field campaigns

The entire field of paleo climate reconstruction relies on natural archives recording climatic fluctuations. Sampling these archives always requires field work. In my case the archives were high alpine glaciers. Three times, I was given the opportunity to participate in such an ice core drilling campaign.

2.1.1 Deep drilling on Fiescherhorn, December 2002

After two months of waiting for stable weather conditions, a deep drilling campaign organized by Margit Schwikowski started on Fiescherhorn glacier in December 2002 (46°33'N, 8°4'E). Previously, several unsuccessful attempts had been made to drill to bedrock on this glacier. To access the drilling site, the team went from Grindelwald to Jungfraujoch and spent one night at the scientific research station there.

From Jungfraujoch a helicopter was used to reach Fiescherhorn glacier. A load of more than 1,000 kg (drilling- & outdoor equipment, equipment for glaciological measurements and food) was flown there. Figure 2.1 shows the upper part of Fiescherhorn glacier, the camp is visible and the snow track leads to the drilling tent. Due to the shortness of daylight, drilling was limited to about 8 hours per day. Although it would have been possible to drill during the night, too, this always seemed to cause problems (the drill got stuck several times). Environmental conditions were rather harsh: During the day, air temperatures never exceeded -7°C and during the night temperature below -25°C were recorded in the sleeping tents. Bedrock was reached after one week of drilling, and the samples were successfully flown down to the valley.

Core processing started in January 2003 by Anne Palmer and was continued by Theo Jenk from summer 2004 onwards. The position of Fiescherhorn glacier is marked in Figure 4.1 on page 73.

Figure 2.1: Drilling on Fieschergletscher in December 2002. The summit on the right is Finsteraarhorn, 4274 m a.s.l. Photographer: Aurel Schwerzmann, ©2002.

2.1.2 The reconnaissance expedition to South America, January 2003

In order to find a suitable glacier archive recording past variations of the El Niño-Southern Oscillation, a field campaign to sub-tropical South America (south of the Atacama desert) was launched in January 2003. The goal was to recover shallow (short) cores from two glaciers such that their suitability could be tested. The trip had been organized by Margit Schwikowski and was funded mainly by Paul Scherrer Institut, Switzerland. Austral summer (December, January, February) is the best season for field work in this region, because the anticylclonic system over the southern Pacific is strong and normally ensures stable weather conditions. The Swiss team included four persons, that flew to Santiago de Chile: Margit Schwikowski (leader), Beat Rufibach (mountain guide) and Theo Jenk and myself (PhD students).

The first part of the mission was devoted to a glacier on a mountain near the capital of Chile, Cerro del Plomo (*Plomo* is the spanish word for *lead*), see Figure 2.2. Although access by car was possible up to the skiing area *La*

Figure 2.2: Cerro del Plomo from camp II. Glaciar Esmeralda is located behind the crest on the left side of the summit.

Parva, from that point onwards, mules had to be used for the transportation of the food and the camping equipment. However, the mules could only reach camp II (4,200 m a.s.l.), from whereon the terrain was too steep for them. The material had then to be carried by the drilling team to camp III (5,100 m a.s.l.) and further on to the glacier. On the glacier (Glaciar Esmeralda) a 5.5 m firn core (consisting of 9 individual segments) was recovered at an altitude of 5,300 m a.s.l. (Figure 2.3). The samples were carried down, first by the team and then by the mules from camp II onwards. Dry ice had been carried up to camp II so that it could be used to cool the samples and prevent them from melting during the transport. Finally, the core segments were shipped to Switzerland in a frozen condition. In total, this part of the campaign lasted for eight days. More technical information about this campaign is given in Section 3.2.1.

After a few days, the team went on to the next glacier that was thought to be an interesting candidate: Glaciar La Ollada on Cerro Mercedario, Argentina (Figure 2.4). This site was even more remote. It takes one day by car to reach the site from the nearest cities Mendoza or San Juan, respectively. 4WD Vans could take up the material only to an altitude of 2,100 m a.s.l., a place called *El Molle* and the track to get there was in poor condition. From that point on, the transport of the equipment was taken over by mules

Figure 2.3: Shallow core drilling on Glaciar Esmeralda, Cerro Plomo.

who had to overcome a difference in altitude of almost 4,000 m to reach the highest point theoretically accessible to them (5,900 m a.s.l.). However, at an altitude of 4,200 m a.s.l., the mules could not continue because of the presence of a steep snowfield (Cuesta Blanca), which they were not able to overcome. From that point on, all the material had to be carried by the participants, themselves.

Because of the long distance and in order to acclimatize, four camps were held before the drilling site was reached. On the glacier, radar measurements were performed, stakes were left on the glacier (for snow accumulation measurements) and a shallow firn core was drilled (13.5 meters, 20 core segments) at an altitude of 6,100 m a.s.l. The next day, the samples were carried down to 4,200 m early one morning, meanwhile the mules had carried up dry ice to ensure that the precious snow remained in a frozen state. The terrain was relatively steep which resulted in rough transport of the material by the mules. When the animals reached the base camp at 2,100 m, it was discovered that some of the segments had been badly damaged. Again, the core segments were shipped to Switzerland in a frozen condition. This part of the campaign took ten days.

The analysis of the two firn cores drilled in January, 2003, showed that Glaciar La Ollada on Cerro Mercedario was a suitable glacier archive recording climate fluctuations of the past (see Chapter 3). Contrastingly, Glaciar Esmeralda on Cerro del Plomo was strongly influenced by melt water formation and percolation destroying its chemical and its isotopic signature. Therefore, the Chilean glacier was regarded unsuitable as a climatic archive. Thus, a deep drilling campaign was planed for February, 2004, attempting

Figure 2.4: Mercedario, the summit in the center, is one of the highest points of the Andes mountain chain. An amplified view of the mountain is inserted. Glaciar La Ollada is located right of the summit and marked by "x".

the recovery of a bedrock core on La Ollada, Argentina, within a time frame of four weeks.

2.1.3 Mercedario field campaign 2004

The participating Swiss team was the same as in 2003 and it was additionally reinforced by the glaciologist Martin Lüthi. This time, the starting point was Mendoza, Argentina. On the way from the city to Barreal (the last village before entering the mountain wilderness) the team was for the first time confronted with what was going to cause great concern later. Although the weather conditions should be most stable during this season of the year, intense thunderstorms and heavy rain prevailed during the car trip. The impact of this weather became obvious the next day when the team tried to move on from Barreal to the base camp. A poor track lead into the mountains and as a consequence of the thunderstorm the day before, part of this track had been buried by mud and stones from a land slide, as illustrated in Figure 2.5 Two days later the track was cleared and the trip could be continued. Although mud had also covered parts of the track higher up, the equipment could still be taken up by car almost to the highest accessible point (2,100 m a.s.l.). From there on, the transport of the material (about 1,000 kg) was taken over by mules. Unfortunately, the period of unstable weather conditions had not come to an end yet. It snowed several times which did not effect the movement of the team but the transport of the drilling equipment was delayed by several days. It was almost impossible to guide the mules through the snow, as the animals were not at all used to that. Finally, they could carry it up to the highest point at the rim of the glacier but only reached a site at 5,700 m a.s.l. near the glacier tongue, where a new camp

Figure 2.5: Road buried by stones and mud from a landslide after a heavy thunderstorm the day before.

was built up. It was intended to carry up the equipment to the drilling site at 6,100 m a.s.l. by the team members, who had already successfully set up the drilling tent. Although this difference in altitude between the last camp and the drilling site does not sound like much, one person could only make it once per day due to the great horizontal distance and due to exhausting conditions at high altitude after a lot of fresh snow had fallen. All that went on too slowly and although drilling could have been started it could not be guaranteed that sufficient time would be left for a safe descent and transport of the material (if drilling had been successful an additional charge of $\sim$400 kg of ice would have had to be taken down). Hence, it was decided to halt the campaign. All the equipment was taken down again and after spending one week in San Juan, Argentina, the drilling team returned to Switzerland.

The scientific work that was completed successfully included the estimation of the accumulation rates at the stakes that had been placed the year before and the determination of their precise positions by differential GPS. Moreover, a meteorological station was run for three days at the camp and recorded insulation, temperature, humidity and wind speed.

2.1.4 Consequences of the field campaign 2004 failure

The failure of this field campaign had personal consequences for myself and for the continuation of my scientific work. The ice core from Mercedario should have provided the samples I thought I was going to base my PhD thesis on. I therefore had to change the subject of my dissertation. Two months after returning to Switzerland and many discussions later I decided to accept Heinz Gäggeler's offer to continue my work on an ice core that had been drilled in September, 2003, on Colle Gnifetti, Switzerland.

This was, of course a quite dramatic change of the subjects switching from climate of subtropical South America under the influence of the El Niño Southern Oscillation to climate reconstruction in central Europe based on an ice core. However, I felt that this ice core had potential. All the European ice core data published so far did not provide isotope records going further back in time than the early 21st century, where stable isotopes are climatically the most important parameter that can be measured in an ice core. We knew from previous studies that this ice core would probably go far back in time, at least 2,000 years (Lüthi and Funk, 2000), so I thought it would be a real challenge.

2.2 Ice cutting

All the ice cores were processed in the cold room at Paul Scherrer Institut at -20°C. The goal of the core processing was to provide samples which were free of any kind of contamination. When the ice cores were recovered on the glacier and taken out of the drill's barrel, they had the shape of a cylinder. The material was relatively clean to begin with but contamination of the surface could not be excluded, first, because the surface of the core segments was in contact with the drill's barrel and second, contamination might also have occurred during packing and sample transport although greatest care was taken.

In order to achieve the goal of preparing decontaminated samples, the outer surface of the samples can be removed completely in the course of the band saw cutting process or samples that still contain part of the original surface can be rinsed with ultra-clean water (18 MΩ cm quality). Although the second method benefits from the fact that less material is needed, its main disadvantage is its limitation to ice. Firn cannot be washed because of its high porosity. It would soak up a large portion of the water (biasing ionic concentrations), especially if this washing procedure were applied on low-density firn ($\rho \leq 0.5\ kg/dm^3$) it would even partly melt away.

Thus, the first procedure was chosen: The removal of the core segment's surface. This was achieved by using a modified commercial band saw operated in the cold room (It can also be achieved by chiselling off the surface manually). The saw had been upgraded by the use of stainless-steel blades, tabletop and saw guides covered with Teflon (Eichler et al., 2000). Before the saw was used, it had to be carefully cleaned each time (especially the blade) such that contamination could be avoided. For the Colle Gnifetti ice core, all cuts but the final sample preparation were lengthwise through the core by pressing the segment onto the saw's guide and pushing it towards the blade. In the first step the cylinder (diameter 8.0–8.2 cm) was cut into two nearly

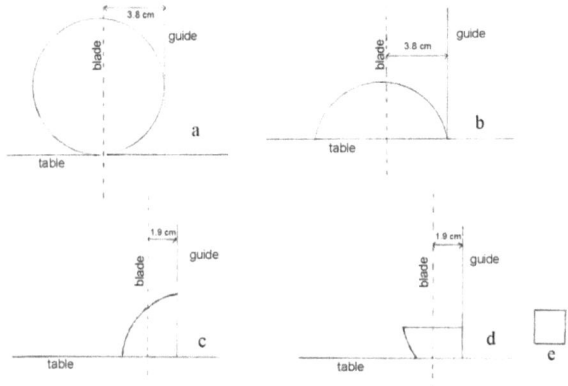

Figure 2.6: Scheme for ice cutting applied for the Colle Gnifetti ice core.

equal slices (see Fig. 2.6, a). The slice on the right side of the blade (3.8 cm) was turned counterclockwise such that the fresh and flat surface would be lay on the table. The next slice was cut at 3.8 cm again (Fig. 2.6, b) and rotated again counterclockwise. The outer parts of the obtained slice were removed by twice cutting at 1.9 cm (Fig. 2.6, c & d). Finally, after cutting four times through the segment a cuboid was obtained whose base had the dimension of 1.9 cm by 1.9 cm (Fig. 2.6, e), the length of the cuboid would correspond to the length of the core segment.

Although many ways exist to achieve the final cuboid (e.g. Knüsel, 2003, p. 30), the method presented here has two main advantages: The segments are guided and therefore touched on the left hand side of the blade. By always rotating the remaining piece counterclockwise it was guaranteed that it was always touched on a surface that would be removed later on. Secondly, by cutting twice at 3.8 cm and 1.9 cm, respectively, the saw only had to be adjusted twice but four consecutive cuts could be made.

The last step was the final sample preparation, cutting the cuboid into individual samples of a typical size varying between 1.5 cm and 7.0 cm according to the desired spatial resolution. Cutting samples smaller than 1.5 cm proved not to be feasible as too much material would be lost due to the thickness of the blade and because the reduction in volume would increase the blank contamination. Only in this last step, the sampled material had to be touched directly by hand on the "fresh" surface. The samples were transferred into containers made of polypropylene that had been precleaned by storing them in regularly changed, ultra-pure water. The samples were stored in the containers until they were analyzed. During the entire sample preparation clean plastic gloves were worn and changed between the respective steps.

2.3 Ion chromatography

Before the samples were analyzed, the sample containers were flushed with nitrogen for 30 seconds such that the laboratory air inside the containers was replaced by clean N_2. This helped reducing contamination by volatile components, especially organic acids like formic acid (HCOOH) and acetic acid (CH_3COOH) but also ammonia (NH_3), present in the laboratory air. These trace gases would otherwise be analyzed as formate, acetate and ammonium (NH_4^+). In the next step the samples were melted and a volume of 900 μL was filled into vials for anions, cations and stable isotope analyses (see Section 2.4), respectively. The vials had been precleaned by storing them in ultra-pure water that had been renewed five times. DIONEX columns were used for the anion and the cation separation, respectively.

Anions (fluoride, acetate, formate, methane sulphonate, chloride, nitrate, sulphate and oxalate) were analyzed on a DIONEX system (DX-500), that had been upgraded with an eluent generator and a self-regenerating trap column. Details for this system are given in Table 2.1. For the eluent preparation, ultra-pure water was first degassed under vacuum and a flow of helium and then used as an eluent. KOH was added to the water by the eluent generator such that the final concentration applied on the column varied between 0.2 mM and 24 mM KOH (gradient separation). The flow rate was 0.5 ml per minute and the recording time for a chromatogram was 15 minutes. Ionic concentrations were measured in a dynamic range from 1 to 2,000 ppb (parts per billion or μg/kg). The concentration of sulphate rarely exceeded this upper limit, then the samples had to be diluted.

Quantitation limits (QL) were calculated based on the noise of the base line and defined as the concentration corresponding to 10 times the standard

Table 2.1: Components of the Dionex anion chromatograph.

Component	Article
Analytical column	IonPac AS11-2mm
Guard column	IonPac AG11-2mm
Pump	Gradient pump GP40
Suppressor	AMMS III-2mm,
Suppression	0.25 mN H_2SO_4
Eluent	KOH, 1-24 mM
Eluent Generator	EG40
Trap column	CR-ATC, self-regen.
Dynamic range	1–2,000 μg/kg
Autosampler	AS 3500
Software	Chromeleon 6.5

deviation (σ) of the base line:

$$QL = 10 \times \frac{\sigma}{S}$$

S: slope of the calibration curve

Quantitation limits correspond to roughly 3 times the detection limit.

Because the base line is not completely flat in any section of the chromatogram, the trend had to be removed for the relatively flat section between 1.0 and 2.0 minutes. The standard deviation was 0.000080μS (micro Siemens). Depending on the different average height of the peaks observed in a multi-element standard (5μg/kg for fluoride and methane sulfonate, 20μg/kg for acetate, formate and oxalate and 50μg/kg for chloride, sulphate and nitrate), different quantitation limits were obtained and listed in Table 2.2. Acetate peak height was low and instable, this is why a range is specified here.

Table 2.2: Quantitation limits for anion chromatography in μg/kg.

F^-	Ac^-	Fo^-	MSA^-	Cl^-	SO_4^{2-}	Ox^{2-}	NO_3^-
0.01	0.08–0.92	0.07	0.08	0.01	0.02	0.04	0.03

Cations (sodium, ammonium, potassium, magnesium and calcium) were analyzed on a Sykam system, details for this system are given in Table 2.3.

The eluent was 23.6 mN sulphuric acid. It was prepared by adding 3.3 mL of concentrated H_2SO_4 to 5 L of ultra-pure water. An isocratic elution was applied at a flow rate of 1.0 mL per minute.

Table 2.3: Components of the Sykam cation chromatograph.

Component	Article
Analytical column	IonPac CG12-4mm
Guard column	IonPac CG12-4mm
Pump	Sykam S1100
Suppressor	CSRS ULTRA II 4mm
Suppression	electrochemical
Eluent	H_2SO_4, 23.6 mN
Eluent Generator	—
Trap column	—
Dynamic range	1–1000 µg/kg
Autosampler	Talbot ASI-5
Software	Chromeleon 6.5

The quantification of ionic concentration was carried out by using multi-ion standard solution in a concentration range of 5–1,000 µg/kg for cations and 2–2,000 µg/kg for anions, respectively. Quadratic calibration curves were used.

Quantitation limits were calculated based on the noise of the base line and defined as 10 times the standard deviation of the base line. Because the base line is not completely flat in any section of the chromatogram, the trend had to be removed for the relatively flat section between 5.5 and 7.0 minutes. The standard deviation was 0.000095µS. Depending on the different average height of the peaks observed in standard with 5 µg/kg, different quantitation limits were obtained and listed in Table 2.4. Generally higher limits than for anions were observed due to lower peak height, i.e. lower signal/noise.

Table 2.4: Quantitation limits for cation chromatography in µg/kg.

Na^+	NH_4^+	K^+	Mg^{2+}	Ca^{2+}
0.3	0.3	0.7	0.4	0.5

2.3.1 Blanks

In order to estimate the blank for the sample preparation procedure, blank ice (frozen ultra-pure water) was cut and analyzed in the same way samples were treated. Table 2.5 shows typical concentrations that were found in the blank ice. For data processing the difference between the obtained

Table 2.5: Typical concentrations found in blank ice (μg/kg). "<QL" means lower than quantitation limit.

Na	NH$_4$	K	Mg	Ca	F	Ac	Fo	MSA	Cl	SO$_4$	Ox	NO$_3$
0.8	1.3	<QL	0.4	3.6	<QL	<QL	2.1	<QL	1.7	0.6	<QL	0.1

ionic concentrations in the blank ice and the concentrations found in freshly prepared ultra-pure water was subtracted from the concentrations found in the ice samples. This way, the data was corrected for the contamination due to the sample preparation. Generally, the maximum values found in the ice cores were up to 3–4 orders of magnitude higher than blank ice concentrations whereas minimum values were in the same order of magnitude but still higher than blank ice concentrations.

2.4 Stable isotope mass spectrometry

The same samples that had been analyzed for major ions were also analyzed for stable isotopes in water: ^{16}O, ^{18}O, ^{1}H and ^{2}H (the latter one is also referred to as D, deuterium). Deviations of the isotopic ratio ($\frac{^{18}O}{^{16}O}$ and $\frac{^{2}H}{^{1}H}$) from *Vienna Standard Mean Ocean Water (VSMOW)* are referred to as δ^{18}O and δD. VSMOW is an artificial standard, prepared and distributed by the IAEA (International Atomic Energy Agency) (Clark and Fritz, 1997, p. 7). For oxygen the isotopic composition is (Baertschi, 1976):

$$\left(\frac{^{18}O}{^{16}O}\right)_{VSMOW} = (2005.2 \pm 0.45) \times 10^{-6}$$

And for hydrogen it is (Hagemann et al., 1970):

$$\left(\frac{^{2}H}{^{1}H}\right)_{VSMOW} = (155.76 \pm 0.05) \times 10^{-6}$$

δ^{18}O and δD are defined as the per-mil (‰) difference between the sample composition and the standard VSMOW (where r stands for isotopic ratio):

$$r_{sample} = \frac{r_{sample} - r_{standard}}{r_{standard}} \times 1000 \text{ ‰}$$

Analyses were carried out by pyrolysis of the samples at 1450°C in a glassy carbon reactor, to produce carbon monoxide ($H_2O + C \rightarrow CO + H_2$) (e.g. Gehre and Strauch, 2003; Saurer et al., 1998). The subsequent measurement of the relative proportions of $^{12}C^{16}O$ and $^{12}C^{18}O$ was done by standard isotope-ratio mass spectrometry. The precision of the measurements was ±0.2‰ for the ice core from Mercedario when a *Delta Plus XL, Finnigan MAT* was used. For samples from Colle Gnifetti the precision was ±0.1‰ after a new generation of the instrument had been set up (*Delta Plus XP, Finnigan MAT*). For δD the precision was not always constant but in the best case ±0.5‰ (worst case: ±1.0‰) was achieved. The deuterium excess, d, was calculated by $d = \delta D - 8 \times \delta^{18}O$ (Dansgaard, 1964). Its precision (estimated by error propagation) varied from 0.9–1.3‰.

Calibration was carried out using an internal, self-made, standard ($\delta^{18}O$=-9.82‰, δD=-70.08‰) that had itself been calibrated against a standard solution distributed by IAEA. The instrument itself already produces absolute values for the isotopic composition such that the standards are in principle only used to correct for any drift during the measurement. Standards of the IAEA are not commercially available and since the distributed quantity is very limited they cannot be used directly for the measurements (R. Siegwolf, personal communication).

2.5 Tritium

Tritium (^{3}H) is a radioactive β emitter ($^3_1H \rightarrow\ ^3_2He + e^-$) and has a half-life of 12.3 years. Since ^{3}H in aqueous form (i.e. HTO) is a volatile compound and since the decay to stable ^{3}He follows emission of a pure β-particle, radioassay of tritium was carried out via liquid scintillation counting (LSC) which is the predominant way for tritium analysis. The counter that was used was a *TriCarb 2770 SLL/BGO, Packard SA, Meriden, Il, USA*: The aqueous sample is mixed with a scintillation liquid and the interaction of the β-particle with the orbital electrons of the target scintillator causes excited states followed by emission of light, which is detected in two photo-multiplier tubes connected in a coincidence circuit. Pulse shape and height analysis in a multi-channel analyzer (MCA) provides a full energy spectrum, which allows distinguishing between radionuclides that emit β-radiation of different energies. A counting window from 0 to 8 keV was taken for weak β-emitting HTO and a counting time of 1000 min was chosen for each sample in order to minimize the measurement uncertainty. The sample size was 10 mL. The detection limit of the method was 10 Tritium Units (TU). 1 Tritium Unit is equal to $\frac{1\ ^3H\ atom}{10^{18}\ ^1H\ atoms}$. 1 TU also equals a radioactivity of 0.12 Bq/kg. Therefore, at a typical concentration of 20 TU ~1440 decays were expected for one sample during the counting time. The analysis was done by J. Eikenberg.

2.6 Nuclear dating with lead-210

Lead-210 (^{210}Pb) is a member of the natural decay chain of uranium-238 via gaseous ^{222}Rn, that is constantly emitted from the earth's crust. Lead-210 is attached to aerosols, that undergo dry or wet deposition anywhere including glaciers. Because of its half-life of 22.3 years, it can be used for dating firn and ice up to an age of ~150 years (Gäggeler et al., 1983). Here, the lead-210 activity was indirectly measured by its decay product ^{210}Po, that is an α-emitter.

The chemical preparation of the ice samples was performed by E. Vogel, University of Berne, the α-decay measurements were performed by L. Tobler.

2.7 Radiocarbon dating

For dating the deeper parts of the Colle Gnifetti ice core radiocarbon (^{14}C) results were used. These results were obtained by T. Jenk and are presented in his PhD thesis (in preparation). Carbonaceous micro-particles were obtained from the melted samples, these particles were oxidized to carbon dioxide which was then reduced to graphite. The ratio ^{14}C to ^{12}C was determined by accelerator mass spectrometry and the sample age was calculated. Details of the method are presented in Jenk et al., submitted in 2005.

Chapter 3

Manuscript in press in Annals of Glaciology 43 (2006), 14–22

A first shallow firn core record from Glaciar La Ollada on Cerro Mercedario in the Central Argentinean Andes

DAVID BOLIUS[1], MARGIT SCHWIKOWSKI[1], THEO JENK[1,2], HEINZ W. GÄGGELER[1,2], GINO CASASSA[3], ANDRES RIVERA[3,4]

[1]Paul Scherrer Institut, CH-5232 Villigen PSI, Switzerland
[2]Department of Chemistry and Biochemistry, University of Bern, Freiestrasse 3, CH-3012 Bern, Switzerland
[3]Centro de Estudios Científicos, Av. Prat 514, Valdivia, Chile
[4]Laboratorio de Glaciología, Departamento de Geografía, Universidad de Chile, Marcoleta 250, Santiago, Chile

ABSTRACT. In January 2003, shallow firn cores were recovered from Glaciar Esmeralda on Cerro del Plomo (33°14′S, 70°13′W, 5300 m a.s.l.), Central Chile and from Glaciar La Ollada on Cerro Mercedario (31°58′S, 70°07′W, 6070 m a.s.l.), Argentina in order to find a suitable archive for paleo climate reconstruction in a region strongly influenced by the El NiñoŰSouthern Oscillation. In the area between 28°S and 35°S, the amount of winter precipitation is significantly correlated to the Southern Oscillation Index with higher values during El Niño years. Glaciochemical analysis indicates that the paleo record at Glaciar La Ollada on Cerro Mercedario is well preserved whereas at the Glaciar Esmeralda the record is strongly influenced by melt water formation and percolation. A preliminary dating of the Mercedario core by annual layer counting results in a time span of 17 years (1986–2002), yielding an average annual net accumulation of 0.45 m water equivalent.

3.1 Introduction

Reconstruction of climatic variability on annual, interannual, decadal and millennial time scales based on ice cores retrieved in the South American Andes has a long tradition. The mountain chain is roughly 7,000 km long and has an average height of 4,000 m. Many mountain glaciers exist, some of which have proved to contain valuable climate archives like Quelccaya (Thompson et al., 1985) and Huascarán (Thompson et al., 1995) which was the first glacier outside the polar regions to provide a climate record extending to the late glacial stage. However, most of the ice core sites have been selected within the tropical part of the Andes (10°N Ű 18°S) which lies within the easterly trade wind circulation and thus the moisture comes from the tropical Atlantic Ocean. The region between 18°S and 28°S is very arid (see Fig. 3.1) and only a few small glaciers exist, despite the great elevation of the mountain chain with several peaks exceeding 6,000 meters. There, the

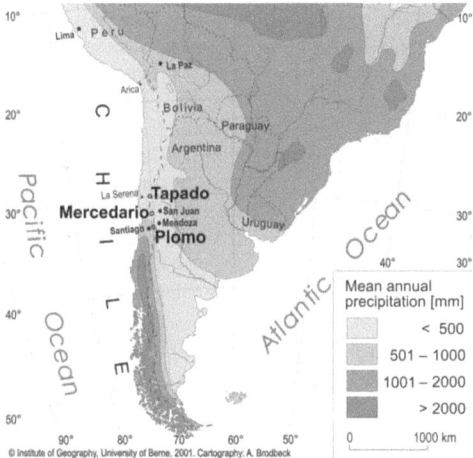

Figure 3.1: Map of southern South-America showing the geographical location of the two drilling sites on Cerro del Plomo and Cerro Mercedario. Mean annual precipitation amount is indicated by the grey scale. The map is adapted from Veit 2000.

formation of glaciers is limited by the low amount of precipitation whereas in most high-mountain areas the limitation is temperature (Kull et al., 2000).

South of 28°S, the climate is increasingly influenced by the Westerlies bringing moisture from the Pacific. Precipitation on the western side of the Andes (e.g., as observed in Santiago de Chile) mainly falls in the austral winter (May to August) when the subtropical anticyclone retreats northwards. The eastern side of the mountain chain is wind-shaded, precipitation is relatively low and falls in the austral summer related to the tropical circulation (e.g. in Mendoza, Argentina).

Although the El Niño-Southern Oscillation (ENSO) influences precipitation over the entire coast of South America, the impact on Andean glaciers is most direct between 28°S and 35°S where the Pacific acts as the moisture source. The amount of winter precipitation in this area is significantly correlated to the Southern Oscillation Index (SOI) (Aceituno, 1988), with higher values during El Niño years. At altitudes between 1380 and 3600 m a.s.l. above average snow accumulation was observed in the Andean sector between 30°S and 35°S when the sea surface temperature anomaly in the Niño 3 region surpassed +1°C during austral winter (May-August) (Escobar and Aceituno, 1998). South of 35°S, it is generally more humid and the influence of ENSO on precipitation declines, becoming anti-correlated in Patagonia, with precipitation decreasing by about 15% during strong El Niño years between 45 and 55°S (Schneider and Gies, 2004). An influence of ENSO on precipitation was also observed in ice cores from tropical South America, but there above average sea surface temperatures across the equatorial Pacific Ocean coincide with lower accumulation rates and/or less negative $\delta^{18}O$ (Bradley et al., 2003; Henderson et al., 1999; Hoffmann et al., 2003; Vuille et al., 2003b,a). The moisture flux of Atlantic origin seems therefore to be remotely controlled by the conditions in the Pacific.

Hence, glaciers between 28°S and 35°S are expected to record a strong ENSO signal and are therefore potential paleoclimatic archives for revealing information about ENSO in the past. The only ice core that has been recovered here was drilled on Cerro Tapado (30°S). Much was learned from this record about sublimation of snow modulating the chemical signal (Ginot et al., 2001a). Variations in $\delta^{18}O$ were ascribed to alternating humid and dry phases related to El Niño events (Ginot et al., 2002). However, the record spanned only a short time period due to a limited glacier thickness (36 m). Therefore the obvious next step was to search for a thicker glacier presumably going further back in time in the geographic vicinity of Cerro Tapado.

Unfortunately, only a few potential ice core sites exist here: Cerro Potro (28°S) is very close to the so-called Şdry axisŤ and strongly influenced by sublimation of snow (complicating the interpretation of records), as indicated by the presence of large penitentes (Ulrich Schotterer, personal communication). Agua Negra glacier (30°S), only a few kilometres away from Cerro

Tapado, is too low in elevation (4600 to 5200 m, Milana and Maturano, 1999), and on Cerro Olivares (30°S, accessible from Agua Negra pass) only a relict glacier exists in the caldera. The study presented here, was carried out to investigate the suitability of two glaciers as climate archives: Glaciar Esmeralda on Cerro del Plomo (33°S, 5300 m, see Fig. 3.1) and Glaciar La Ollada on Cerro Mercedario (32°S, 6070 m). Cerro del Plomo is close to Santiago de Chile and can be reached from the skiing area ŞLa ParvaT̃ within three days walk. Mercedario is about 80 km north of Aconcagua, the highest mountain of the Americas. Access leads through a remote area of Argentina, with Mendoza and San Juan being the nearest cities.

Mean January (austral mid summer) temperature at the nearest high-altitude meteorological station Cristo Redentor (32.82°S, 70.07°W, 3852 m a.s.l.) is 4°C. Using a standard lapse rate (6.5°C/1000 m) a mean January temperature of -6°C and -11°C for Cerro del Plomo (5300 m) and Cerro Mercedario (6070 m a.s.l.), respectively, could be assumed.

3.2 Methods

3.2.1 Drilling campaigns

In December 2000 a first glaciological and chemical survey was performed on Glaciar Esmeralda on Cerro del Plomo (33°14′S, 70°13′W, 5300 m a.s.l., Central Chile). The glacier surface was flat and no ŞpenitentesT̃ (snow sculptures that form under very strong sublimation) were observed. Nevertheless, concentrations of ionic species in snow samples from a pit study showed enrichment in the surface layer attributable to sublimation such that this post-depositional effect has to be taken into account. Radar measurements of the glacier thickness were performed on two profiles on the upper part of the glacier. Overall, Glaciar Esmeralda had seemed to be a suitable candidate for bearing an ice archive.

In January 2003, during an eight-day field campaign, a shallow core was retrieved from Glaciar Esmeralda, using the light-weight, portable drilling device ŞFELICS smallT̃ (Ginot et al., 2001b). Whereas the first four core segments (2.5 m) consisted of homogeneous firn, superimposed ice was encountered from this depth onwards. Drilling was stopped after 5.5 m when 9 core segments had been retrieved. The segments were put into polyethylene bags and carried down to the camp at 4200 m early in the day (temperatures below 0°C). From there on, they were cooled with dry ice until they were shipped to Switzerland in a frozen condition.

A few days after drilling on Cerro del Plomo, Glaciar La Ollada on Cerro

Mercedario (31°58′S, 70°07′W, 6070 m a.s.l., Argentina) was approached. From the highest point accessible by car (2100 m) mules were used for carrying the equipment. The tongue of Glaciar La Ollada extends downwards to an altitude of 5700 m approximately, but no end moraine was obvious. Around the centre of the glacier, radar measurements of the ice thickness were performed on a near-circular traverse. In a flat area where the deepest ice exists, a 13.5 m long shallow core was drilled at an altitude of 6070 m (20 core segments). The firn was compact and homogeneous. Additionally, a set of nine accumulation stakes were mounted on the glacier surface and their position was determined by GPS. The samples were carried to 4200 m elevation early in the morning. From 4200 m, the samples were cooled with dry ice and transported down by mules to the valley. Unfortunately, this rough transport resulted in damage of a few core segments, leading to gaps in the data. The samples arrived in Switzerland in a frozen condition.

In February 2004, a deep drilling campaign was attempted. Due to persistent bad weather conditions, the transport of the drilling equipment up to the glacier was delayed several times and the mules did not reach the highest point that is theoretically accessible to them (5900 m a.s.l.). Consequently the drilling could not be started. Accumulation rates at the stakes and their precise positions by differential GPS could be determined and a meteorological station was run for three days at the lower part of the glacier (5800 m a.s.l.).

3.2.2 Radar survey

The radar system used at both sites was a portable HF impulse-type, applied successfully in glaciers of north-central Chile, Patagonia and Antarctica (Rivera and Casassa, 2002). The transmitter was a Narod model (Narod and Clarke, 1994), powered by a 12 V battery, which generates a pulse of 1100 V, a rise time < 2 ns and a pulse repetition rate of 512 pulses/s. Two 8 m-long half-dipole dipole antennae loaded with resistors were connected to the transmitter, generating an electromagnetic wave with a central frequency of ∼5 MHz. Wires were inserted inside webbing tape, thus protecting the antennae and allowing their use as regular mountain ropes. The receiver consisted of a Tektronics THS-720 digital storage oscilloscope, connected to the receiving antennae (of the same characteristics as the transmitting ones) by means of a balun. Data were stored in the field with a Husky MP 2500 portable PC connected to the oscilloscope through the RS-232 port. Averaging of 16 traces was performed using the oscilloscope, storing one averaged trace every 4 seconds. Surface coordinates were recorded every 1 sec. by differential GPS using two GPS receivers (Trimble Geoexplorer II), with single-frequency C/A code and baselines < 500 m, which gave an accuracy of < 1 m in horizontal coordinates, and < 2 m in elevation.

3.2.3 Chemical analysis

The two shallow cores were processed at -20°C in the cold room. A detailed visual stratigraphy was recorded over the entire length of the two cores. The core segments were weighed, photographed and cut with a band saw, such that the outer part was removed and the inner part, that was contamination-free, was sampled at a spatial resolution of 2.5 cm. Individual samples were transferred into pre-cleaned containers and stored at -20°C. The samples were then melted and analyzed for major ions (Na^+, NH_4^+, K^+, Mg^{2+}, Ca^{2+}, F^-, CH_3COO^-, $HCOO^-$, $CH_3SO_3^-$, Cl^-, NO_3^-, SO_4^{2-}) by ion chromatography using standard procedures (Eichler et al., 2000) as well as for stable oxygen isotopes ($\delta^{18}O$). $\delta^{18}O$ analysis were carried out by pyrolysis of the samples at 1450°C in a glassy carbon reactor, to produce carbon monoxide (CO). The subsequent measurement of the relative proportions of $^{12}C^{16}O$ and $^{12}C^{18}O$ was done by standard isotope ratio-mass spectrometry (Delta Plus XL, Finnigan MAT). $\delta^{18}O$ is defined as the per-mil difference between the sample composition and the Standard Mean Ocean Water, SMOW. The precision of the measurements was ± 0.2‰.

3.3 Results

Two radar profiles with a length of 390 m and 290 m, respectively, were measured at the upper part of Glaciar Esmeralda (Fig. 3.2), at an elevation of ca. 5330 m. They showed a V-shaped (valley) glacier bed, with a maximum un-migrated thickness of 92 m of ice in profile TT' (Fig. 3.2) and 99 m in profile DD' (Fig. 3.2), as shown by a raster display of the radar data in Fig. 3.3. The surface gradient is small in this section of the glacier, with an average slope of 3 % for the diagonal profile (DD' in Fig. 3.2), and 5 % for the transverse profile (TT' in Fig. 3.2). Based on the radar data (e.g. Fig. 3.3) bed gradients are much larger, up to 60%, so that geometrical distortions of the bed echoes are important which can be corrected by 3D migration of the data. 2D migration was performed with a Kirchoff algorithm. As expected, the bed appears as a U-shaped valley in the 2D migrated profiles (not shown), with two-way travel times about 18 % larger than the non-migrated travel times, and migrated ice thickness values about 8 % larger than the non-migrated values. Maximum ice thickness measured at Glaciar Esmeralda is therefore ∼107 m ± 13%. The precision of the radar results

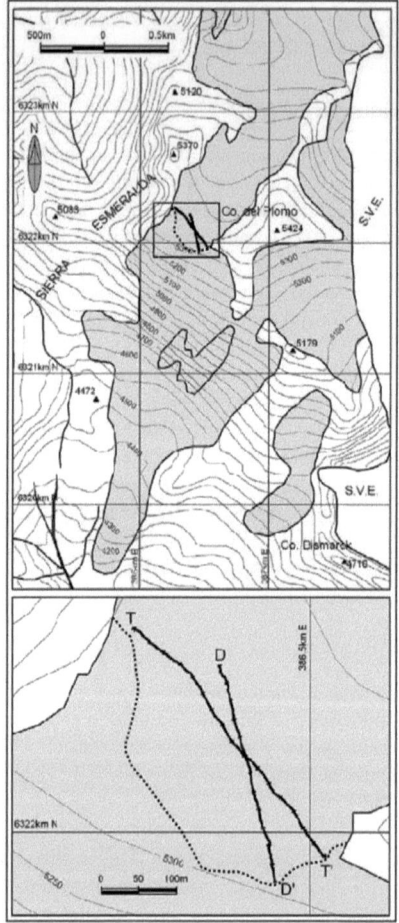

Figure 3.2: Map showing the position of the transverse (TT') and diagonal (DD') radar profiles obtained on Glaciar Esmeralda, Cerro del Plomo in December 2000. Glacier extensions are taken from the 1:50,000 scale map of Instituto Geográfico Militar, and include some areas of rock covered by snow. Glaciar Esmeralda and Glaciar El Plomo are now separated by a rocky ridge at ca. 5350 m, indicated by the broken line. The map is adapted from Rivera et al. (2001).

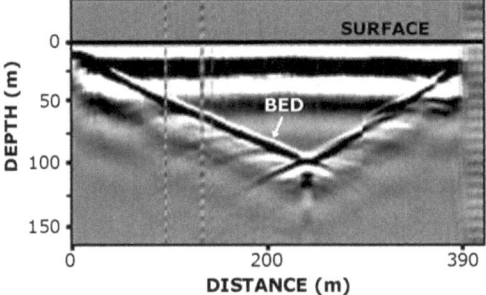

Figure 3.3: Non-migrated raster image of radar data corresponding to the transverse profile TT' of Fig. 3.2. The glacier surface (time zero) appears flat because it has not been corrected using the GPS elevation data. The glacier bed is the V-shaped reflection which appears as a transition from grey to white on the image, with a maximum non-migrated depth of 92 m at the centre of the valley. The figure is adapted from Rivera et al. (2001).

was assessed by comparing the migrated ice thickness at the intersection of both profiles of Fig. 3.3.

The firn density on Glaciar Esmeralda is influenced by the presence of superimposed ice that was found starting at a depth of 2.5 m (Fig. 3.4). The first four core segments consist of firn with low density (<0.5g/cm^3) and show the expected density increase due to compaction. The last three core segments consist only of superimposed ice as indicated by their high density close to 0.9 g/cm^3. Thus, Glaciar Esmeralda is strongly influenced by formation and refreezing of melt water. Concentrations of major ions are affected as well by this post depositional process. They are much lower in the upper 2.5 meters (firn) where percolating water has removed the soluble species (see Fig. 3.5). At the firn-ice boundary, where melt water refroze and formed the superimposed ice, the concentrations suddenly increase as dissolved ions were trapped. Generally, mean concentrations of ionic species are relatively high, compared to concentrations at Cerro Mercedario and Cerro Tapado (Table 3.1). This might reflect the closer vicinity to anthropogenic emission sources such as the city of Santiago, since especially the concentrations of SO_4^{2-}, NO_3^-, and NH_4^+ are elevated. Also the $\delta^{18}O$ record seems to be affected by melt water percolation, indicated by a smoothing of the signal below 1 m water equivalent (weq.) depth (Fig. 3.5).

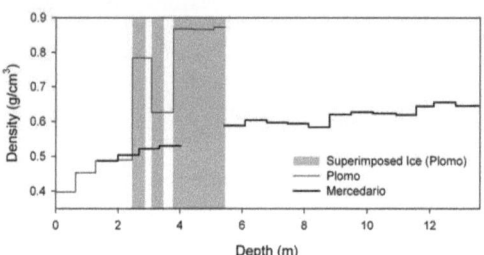

Figure 3.4: Firn density versus depth of the shallow cores from Cerro del Plomo and Cerro Mercedario. Superimposed ice layers in the Plomo core are indicated by grey bars.

Table 3.1: Mean concentrations of major ions and mean $\delta^{18}O$ of the shallow cores from Cerro del Plomo and Cerro Mercedario compared to values at Cerro Tapado (time period 1986–1999).

	Na^+	NH_4^+	K^+	Mg^{2+}	Ca^{2+}	Cl^-	NO_3^-	SO_4^{2-}	$\delta^{18}O$
	µeq./L								‰
Mercedario	2.36	1.07	0.16	0.90	3.96	1.25	2.82	5.14	-21.4
Plomo	1.54	3.09	0.14	1.86	11.24	2.15	4.53	17.28	-18.6
Tapado	1.60	1.17	0.37	0.68	9.76	1.90	3.11	8.55	-18.8

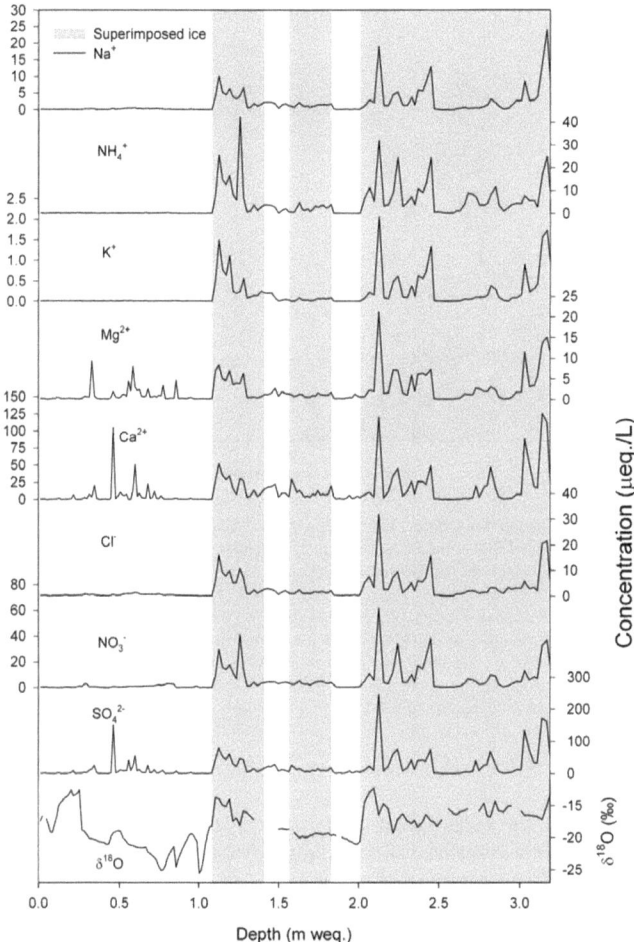

Figure 3.5: Major ion concentration and $\delta^{18}O$ records from Cerro del Plomo. Superimposed ice layers are indicated by grey bars.

At Glaciar La Ollada on Cerro Mercedario (Fig. 3.6) the shallow core was drilled at the deepest part of the glacier with an estimated depth of 137 m ±13% (Fig. 3.7). The increase of density with depth is steady

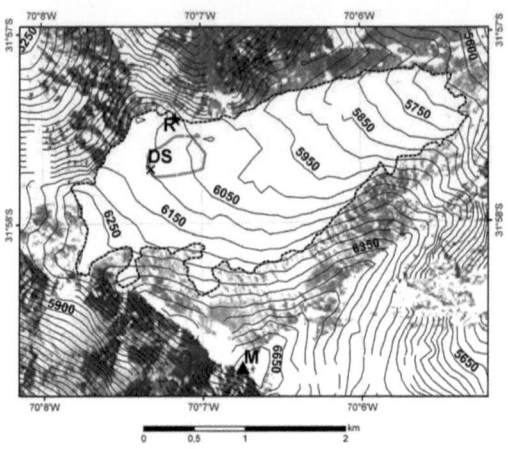

Figure 3.6: Glaciar La Ollada, Cerro Mercedario. Grey-scale composite of bands 1, 2 and 3 of a Landsat-ETM+ image at 14.25 m resolution acquired 28 February 2000 (Earth Science Data Interface at the Global Land Cover Facility, The University of Maryland). Contour lines at 50 m interval were derived with a cubic spline interpolation method using the Shuttle Radar Topography Mission (SRTM) digital elevation model of February 2000 at 90 m horizontal resolution. Elevation data are expressed as ellipsoidal heights. Orthometric or sea level heights for the area are 33 m lower than the ellipsoidal heights according to the EGM96 geoidal model. M is the summit of Mercedario, with a 90-m pixel SRTM ellipsoidal height of 6701 m. R is the base station on rock used as the 2003 reference GPS site. DS is the 2003 drilling site. The 2003 radar profile is shown as a grey near-concentric line.

(0.4 g/cm^3 at the surface, 0.65 g/cm^3 at a depth of 13.5 m, Fig. 3.4). The core consisted of compact firn with only a few minor ice lenses (all thinner than 1 cm). Concentrations of major ions (e.g. SO_4^{2-} or Ca^{2+}) span several orders of magnitude and are strongly correlated to each other (Fig. 3.8). Six layers with visible coarse inorganic particles were found with individual pieces having diameters up to 2 mm. They show up in the major ion record as concentration peaks (Fig. 3.8), since the particles were partially dissolved when the samples were melted. The $\delta^{18}O$ record reveals strong fluctuations ranging from -11.6‰ to -28.3‰, with a mean value of -21.4‰ (Fig. 3.8).

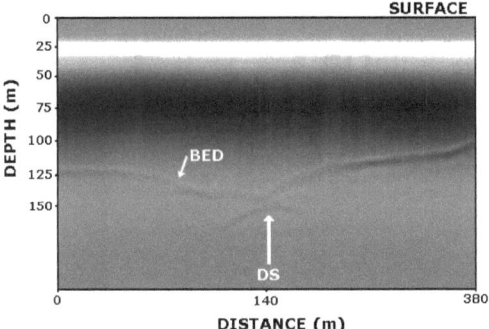

Figure 3.7: Radar profile of the vicinity of the 2003 drilling site (DS) on Glaciar La Ollada, Cerro Mercedario. The topmost edge of the figure corresponds to the ice surface. DS is the deepest point within the profile, with an estimated non-migrated depth of 137±18 m.

Maxima in the $\delta^{18}O$ signal tend to coincide with high concentrations of major ions. This is best documented for the SO_4^{2-} record (Fig. 3.9, page 59).

At three stakes placed within a distance of 100 m from the drilling site a mean accumulation of 78 cm of snow was observed from February 2003 to February 2004, with values ranging from 62 cm to 88 cm of snow. The remaining six stakes could not be found, presumably because they were covered by snow. The observed accumulation is therefore a minimum estimate. Night time temperatures recorded by the meteorological station on the glacier at 5800 m a.s.l. ranged between -12°C and -20°C. Daytime temperatures were not recorded reliably due to technical problems with the station. Relative humidity ranged from 10% to 90%.

3.4 Discussion

3.4.1 Seasonality of precipitation and sublimation

Because of the strong influence of melting on the glaciochemical record, the Glaciar Esmeralda on Cerro del Plomo is considered unsuitable as paleo climate archive. The discussion will therefore focus on the results from the Mercedario core. The fluctuations in the Mercedario $\delta^{18}O$ record are assumed to reflect variations in temperature during the different seasons. Thus, minima

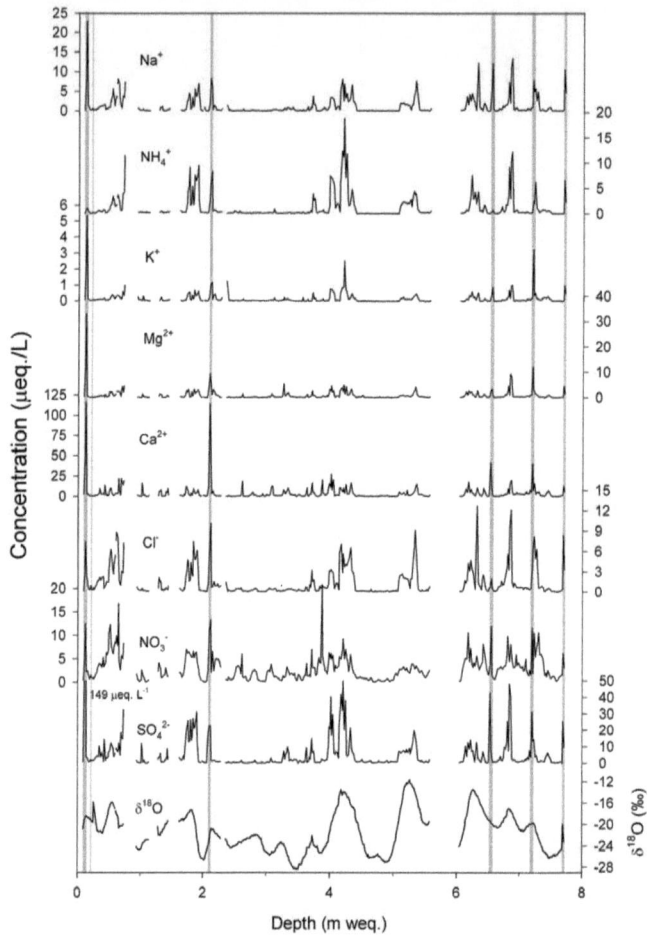

Figure 3.8: Major ion concentration and $\delta^{18}O$ records from Cerro Mercedario. Grey bars indicate the presence of layers with visible coarse particles. Data gaps are due to core damage during transport.

(more negative values) in the $\delta^{18}O$ record are attributed to austral winter precipitation (June, July, August) and maxima to austral summer (December, January, February). Fluctuations of the concentration of major ions match with the $\delta^{18}O$ record such that high concentrations coincide with high values of $\delta^{18}O$ attributable to austral summer precipitation. All major ions show this behaviour as illustrated for SO_4^{2-} (Fig. 3.9). This supports the

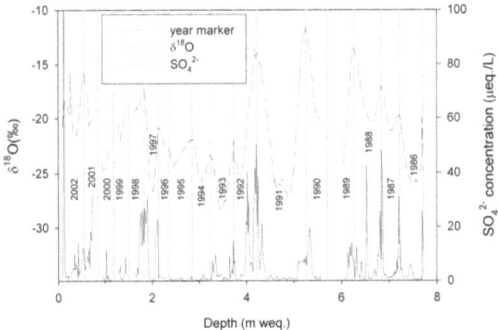

Figure 3.9: Mercedario SO_4^{2-} concentration, $\delta^{18}O$ record and suggested annual layers.

idea of seasonality as the driving force behind the observed fluctuations, as summer snow is expected to show increased ion concentrations compared to winter snow. Such a control of high-altitude aerosol concentrations by the seasonally variable intensity of vertical mixing has been studied intensively in the European Alps (Baltensperger et al., 1997) and the process is assumed to be relevant in this subtropical region, too. This assumption is corroborated by the fact that formation of convective clouds in the afternoon was observed several times in austral summer on Plomo and Mercedario. The second process probably leading to enrichment of chemical species in snow is sublimation. This process is also more important during summer because of high solar insolation and low relative humidity of the air, the two main factors driving sublimation. It was shown experimentally on Cerro Tapado that sublimation of snow is a very important postdepositional process in the subtropical Andes modulating the glaciochemical record (Ginot et al., 2001a). Irreversibly deposited ionic species (e.g. SO_4^{2-} or Cl^-) are enriched in the topmost few centimetres of the surface when the snow evaporates during the dry season. This is very pronounced at the Cerro Tapado, located in the vicinity of the Şdry axisŤ. The effect becomes visible in the chemical record, where several narrow peaks indicate enrichment as illustrated on the example of Cl⁻ in Fig. 3.10. Cl⁻ was selected because this ion is not strongly influenced by dry deposition (as would be the case for Ca^{2+}) nor is it easily released from the snow in a volatile form at this glacier site (Ginot et al., 2001a). In Fig. 3.10 it can be recognized that Cl⁻ peaks in the record from Mercedario are much broader whereas many more ŞspikesŤ exist in the record of Cerro Tapado. This is an indication that sublimation modulates the glaciochemical record more strongly on Cerro Tapado than on Mercedario. This assumption is in agreement with increased humidity and a less pronounced dry season expected at Mercedario, located 150 km to the south of Cerro Tapado and thus farther away from the Şdry axisŤ. The difference in solar radiation between Cerro Mercedario and Cerro Tapado is small due to geographical vicinity and cannot explain the different strength of sublimation at the two sites. The different shape of the peaks cannot be due either to an unequal core cutting resolution which was comparable for both cores.

Winter snowfall was expected to be the dominant source of precipitation in this region as this type of pattern was typical for Cerro Tapado. However, the observed fluctuations at Mercedario suggest that precipitation also occurs in summer. This idea was confirmed when the deep drilling campaign in February 2004 failed as heavy snowfall did not allow the mules to reach the highest camp accessible to them (5900 m a.s.l.). Nevertheless, the major portion of precipitation falls in winter. This is suggested by the $\delta^{18}O$ record showing broader winter minima compared to narrow summer maxima (Fig. 3.9). Accordingly, the $\delta^{18}O$ frequency distribution is skewed towards more negative values. The finding of a less pronounced dry season and a generally more humid atmosphere at Cerro Mercedario compared to Cerro Tapado is in agreement with precipitation data from the City of Santiago de Chile (33°S) and Pisco Elqui (30°S), both at a distance from the coast of about 80 km. The North-South gradient in atmospheric humidity is reflected in small amounts of summer precipitation in Santiago de Chile whereas this season is completely dry in Pisco Elqui (Fig. 3.11). More summer precipitation than in Santiago was found at the higher-elevation sites Punta de Vacas (2416 m a.s.l.), Puente del Inca (2720 m a.s.l.), and Cristo Redentor (3862 m a.s.l.) (Videla, 1997), all three located in the Andean mountain range close to Aconcagua (33°S). The increasing amount of summer precipitation with elevation might be due to lifting of air masses from the West, containing some humidity, or moisture advection from the East. This region forms part of the transition area between trade winds and westerlies. In San Juan, Argentina

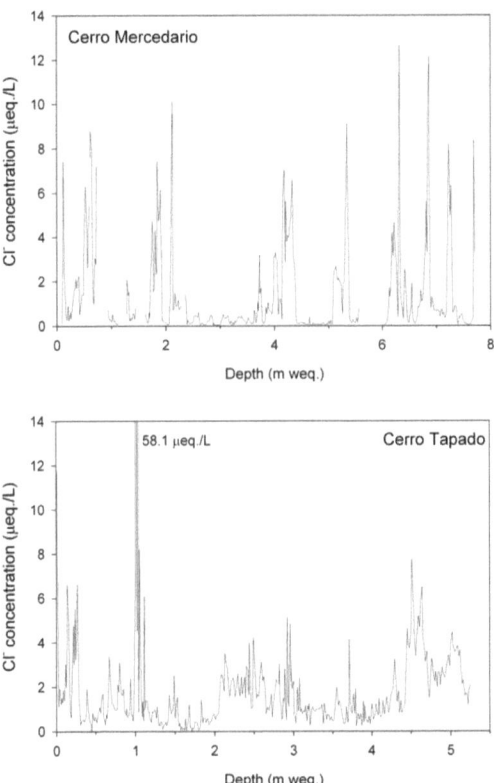

Figure 3.10: Cl⁻ records of Cerro Mercedario (upper panel) and Cerro Tapado (lower panel), representing a comparable time interval.

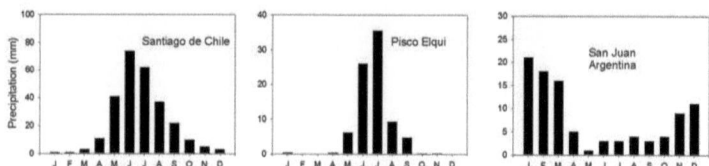

Figure 3.11: Seasonal cycle of precipitation at Santiago de Chile (33°S), Pisco Elqui (30°S), and San Juan, Argentina (32°S). Data are averages for the time period 1961–2003 and were compiled from the Global Historical Climatology Network GHCN, (Vose et al., 1992).

(32°S), situated east of the Andes, on the other hand, the pattern is opposite with maximum amounts in summer, indicating an Atlantic moisture source.

The annual net accumulation rate on Cerro Mercedario (0.45 m weq., see below) is higher than 0.31 m weq. on Cerro Tapado (Ginot et al., 2002). The $\delta^{18}O$ signal in the Cerro Tapado core does not show seasonal variations (Fig. 3.12), which can be explained by the absence of summer accumulation (Ginot et al., 2002). In fact, in the rare case of summer precipitation events on Cerro Tapado, this snow is lost by sublimation. The preservation of winter and summer accumulation at Cerro Mercedario and a smaller influence of post-depositional processes lead to a simpler interpretation of the glaciochemical record resulting in better identification of annual layers.

3.4.2 Preliminary dating and net accumulation

A preliminary dating of the Mercedario core was obtained by annual layer counting using the minima in the $\delta^{18}O$ record as indicated in Fig. 3.9. The resulting time span covered by the core is 17 years (1986 – 2002), yielding an average annual net accumulation of 0.45 m weq. This value is in good agreement with the observed snow accumulation at the three stakes for the time period from February 2003 to February 2004. It turns out that more negative $\delta^{18}O$ values in the Mercedario core seem to occur during known El Niño events (years 1997, 1993, 1991, 1986, Fig. 3.12). Whether this is coincidence or not, and what possible mechanisms could be responsible for such an effect will be subject to future investigations. Above average accumulation in El Niño years was expected to occur on Mercedario, as this was reported for nearby Piloto glacier (Leiva, 1999)(Leiva, 1999) and Echaurren glacier (Escobar et al., 1995)(Escobar and others, 1995) and also for Cerro Tapado.

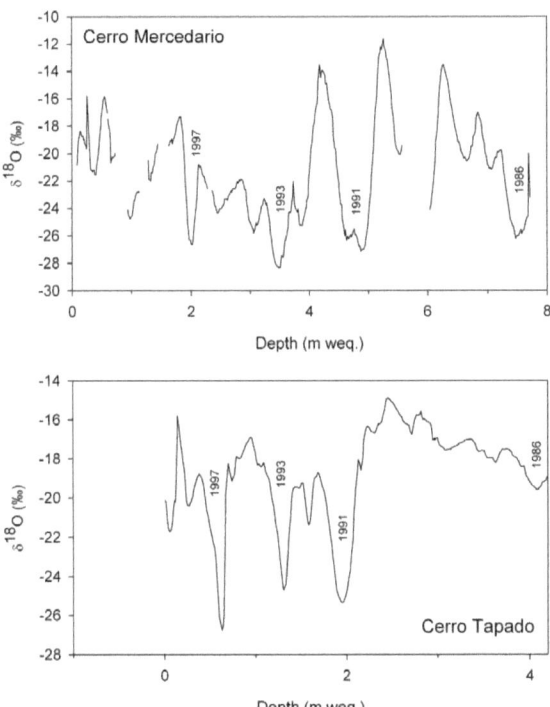

Figure 3.12: $\delta^{18}O$ records of Cerro Mercedario (upper panel) and Cerro Tapado (lower panel), representing a comparable time interval.

At Cerro Mercedario this is particularly obvious for the years 1986 and 1991. The dating relies only on annual layer counting based on fluctuations in the $\delta^{18}O$ and the major ion concentration records and must therefore be regarded as preliminary. Moreover, the layer counting is hampered because of the gaps in the records due to damage of firn core material during transportation. Independent dating methods like ^{210}Pb-dating (Crozaz and Langway, 1966; Gäggeler et al., 1983) or the use of the tritium activity maximum in 1964 as a reference horizon could not be applied in this study due to due the limited time span of the core.

3.4.3 Influence of melt water formation and percolation

As discussed above, huge layers of superimposed ice were found in the core from Cerro del Plomo. They indicate that the glacier is strongly influenced by melt water formation and percolation. In the Cerro Mercedario core, on the other hand, very few and thin ice layers were found. Thus, melt water formation is not an important post-depositional process here. Most of the ice lenses match with maxima in the $\delta^{18}O$ record of Cerro Mercedario. This suggests that melt water, which is produced in austral summer when air temperatures are highest, does not percolate deeply before it refreezes. Thus, the drilling site on Cerro Mercedario is located in the upper percolation zone near the dry-snow line (Paterson, 1994), indicating a very low mean annual air temperature.

The strong discrepancy in melt water formation is attributed to the different altitudes of the two sites with La Ollada on Cerro Mercedario being 800 m higher. Originally, mean summer temperatures of -6°C and -11°C for Glaciar Esmeralda (5300 m) and Glaciar La Ollada (6070 m), respectively, based on temperature extrapolation (6.5°C/1000 m) from the nearest high-altitude meteorological station, Cristo Redentor (32.82°S, 70.07°W, 3852 m a.s.l.) were expected. The intense melting observed on Glaciar Esmeralda is in disagreement with the estimated summer temperatures of -6°C. However, extrapolation is difficult, as there is a strong north-to-south gradient in that area. At Cerro Tapado, located 150 km further north, mean summer temperatures (recorded from December 15, 1998 to February 15, 1999) were -2.8°C at 5260m and +5.0°C at 4215 m (Ginot et al., 2001a). Mean January temperatures of +4°C at 4720 m were measured at the Agua Negra pass (30°S) (Schrott, 1994) and strong melting was observed on Glaciar Agua Negra, which extends between 5200 and 4600m a.s.l. (Milana and Maturano, 1999).

A puzzling stratigraphic feature of the core from Glaciar La Ollada are the layers of visible coarse particles. These particles must originate from snow/ice-free slopes around the glacier. However, the glacier surface is relatively flat and the drilling site is more than 400 m away from the margins of the glacier. Thus, several prerequisites exist for their formation: strong storms blowing coarse particles over the glacier are needed, the glacier surface must be hard to allow for particle saltation and suspension (Nishimura and Hunt, 2000), and the edges of the glacier need to be snow/ice free (late summer). Hence, extreme conditions seem to occur from time to time on Mercedario.

3.5 Conclusion

The well-preserved glaciochemical signal at Glaciar La Ollada on Cerro Mercedario is very promising for obtaining a longer time record. A new core was drilled to a depth of 104 m in February 2005. This record should provide high-resolution paleo climate data from a region strongly influenced by the El NiñoŰSouthern Oscillation in such a way that El Niño years coincide with higher values of winter precipitation. ENSO related climate variations of the past in this region are poorly documented due to the scarcity of suitable archives. However, it cannot yet be estimated how far back in time the new record is going to provide information. Given the estimated annual net accumulation rate of 0.45 m weq. together with a core length of 104 m, a several-century record should be contained. An upper limit for the age at bedrock can hardly be given as the thinning rate is difficult to predict. An ice core from Quelccaya ice cap in Peru (Thompson et al., 1985) was slightly longer (163.6 m) and the ice at bedrock was 1500 years old (the annual layer thickness was 1.2 m ice equivalent at the surface). Another core from Col de Huascarán was 166.1 m long, with an annual layer thickness of around 1.5 m ice eq. and ice dating back to the late glacial stage (dated 19,000 B.P.) (Thompson et al., 1995). The core from Illimani, Bolivia, with an annual accumulation rate of 0.58 m weq. and a length of 138.7 m covered approximately the last 18,000 years (Knüsel et al., 2003; Ramirez et al., 2003). The finding that the Esmeralda glacier on Cerro del Plomo, even at an altitude of 5300 m in the subtropics, is strongly influenced by melt water formation and percolation underlines again how threatened these valuable archives are due to current global warming and how important their investigation is before they become unsuitable or disappear.

ACKNOWLEDGEMENTS

Financial support from the Swiss National Science Foundation (Grant no.

200021-100289) is highly acknowledged. Thanks go to Beat Rufibach for drilling, Felipe Contreras, Gabriel Cabrera, and Alejandro Mirando for support in the field as well as to Matthias Saurer, Rolf Siegwolf and Daniel Theiss for helping with the isotope ratio mass spectrometer. FONDECYT Project 1040989 partially supported GC.

Manuscript[1] reprinted from the Annals of Glaciology with permission of the International Glaciological Society.

---

[1] The manuscript is available for download: http://www.igsoc.org/annals/43/a43a009.pdf (May, 2010)

Bibliography

Aceituno, P., 1988. On the functioning of the Southern Oscillation in the South American sector. *Monthly Weather Review*, **116(3)**, 505–524.

Baltensperger, U., Gäggeler, H., Jost, D., Lugauer, M., Schwikowski, M., Weingartner, E., and Seibert, P., 1997. Aerosol climatology at the high-alpine site Jungfraujoch, Switzerland. *Journal of Geophysical Research - Atmospheres*, **102(D16)**, 19 707–19 715.

Bradley, R., Vuille, M., Hardy, D., and Thompson, L., 2003. Low latitude ice cores record Pacific sea surface temperatures. *Geophysical Research Letters*, **30(4)**.

Crozaz, G. and Langway, C., 1966. Dating Greenland firn-ice cores with ^{210}Pb. *Earth and Planetary Science Letters*, **1(4)**, 194–196.

Eichler, A., Schwikowski, M., Gäggeler, H., Furrer, V., Synal, H., Beer, J., Saurer, M., and Funk, M., 2000. Glaciochemical dating of an ice core from upper Grenzgletscher (4200 ma.s.l.). *Journal of Glaciology*, **46(154)**, 507–515.

Escobar, F. and Aceituno, P., 1998. Influencia del fenómeno ENSO sobre la precipitación nival en el sector andino de Chile Central, durante el invierno austral. *Bulletin de l'Institut Français d'Études Andines*, **27**, 753–759.

Escobar, F., Casassa, G., and Veronica, P., 1995. Variaciones de un glaciar de montaña en los Andes de Chile central en las ultimas dos decadas. *Bulletin de l'Institut Français d'Études Andines*, **24(3)**, 683–695.

Gäggeler, H., von Gunten, H., Rössler, E., Öschger, H., and Schotterer, U., 1983. ^{210}Pb Dating of cold alpine firn/ice cores from Colle Gnifetti, Switzerland. *Journal of Glaciology*, **29(101)**, 165–177.

Ginot, P., Kull, C., Schwikowski, M., Schotterer, U., and Gäggeler, H., 2001a. Effects of postdepositional processes on snow composition of a subtropical glacier (Cerro Tapado, Chilean Andes). *Journal of Geophysical Research - Atmospheres*, **106(D23)**, 32 375–32 386.

Ginot, P., Schwikowski, M., Schotterer, U., Stichler, W., Gäggeler, H., Francou, B., Gallaire, R., and Pouyaud, B., 2002. Potential for climate variability reconstruction from Andean glaciochemical records. *Annals of Glaciology*, **35**, 443–450.

Ginot, P., Stampfli, D., Stampfli, F., Schwikowski, M., and Gäggeler, H., 2001b. FELICS, a new ice core drilling system for high-altitude glaciers. *Memoirs of National Institute of Polar Research*, **56(Special Issue)**, 38–48.

Henderson, K., Thompson, L., and Lin, P., 1999. Recording of El Niño in ice core $\delta^{18}O$ records from Nevado Huascarán, Peru. *Journal of Geophysical Research - Atmospheres*, **104(D24)**, 31 053–31 065.

Hoffmann, G., Ramirez, E., Taupin, J., Francou, B., Ribstein, P., Delmas, R., Durr, H., Gallaire, R., Simões, J., Schotterer, U., Stievenard, M., and Werner, M., 2003. Coherent isotope history of Andean ice cores over the last century. *Geophysical Research Letters*, **30(4)**.

Knüsel, S., Ginot, P., Schotterer, U., Schwikowski, M., Gäggeler, H., Francou, B., Petit, J., Simões, J., and Taupin, J., 2003. Dating of two nearby ice cores from the Illimani, Bolivia. *Journal of Geophysical Research - Atmospheres*, **108(D6)**.

Kull, C., Grosjean, M., and Veit, H., 2000. Late Pleistocene climate conditions in the North Chilean Andes drawn from a climate-glacier model. *Journal of Glaciology*, **46**, 622–632.

Leiva, J., 1999. Recent fluctuations of the Argentinian glaciers. *Global and Planetary Change*, **22(1-4)**, 169–177.

Milana, J. and Maturano, A., 1999. Application of Radio Echo Sounding at the arid Andes of Argentina: the Agua Negra Glacier. *Global and Planetary Change*, **22(1-4)**, 179–191.

Narod, B. and Clarke, G., 1994. Miniature high-power impulse transmitter for radio-echo sounding. *Journal of Glaciology*, **40(134)**, 190–194.

Nishimura, K. and Hunt, J., 2000. Saltation and incipient suspension above a flat particle bed below a turbulent boundary layer. *Journal of Fluid Mechanics*, **417**, 77–102.

Paterson, W., 1994. *The Physics of Glaciers*. Pergamon/Elsevier, Oxford, third edition.

Ramirez, E., Hoffmann, G., Taupin, J., Francou, B., Ribstein, P., Caillon, N., Ferron, F., Landais, A., Petit, J., Pouyaud, B., Schotterer, U., Simões, J., and Stievenard, M., 2003. A new Andean deep ice core from Nevado Illimani (6350 m), Bolivia. *Earth and Planetary Science Letters*, **212(3-4)**, 337–350.

Rivera, A. and Casassa, G., 2002. Detection of ice thickness using radio-echo sounding on the Southern Patagonia Icefield. In: The Patagonian Icefields: A Unique Natural Laboratory for Environmental and Climate Change Studies. Eds. G. Casassa, F. Sepúlveda and R. Sinclair. Series of the Centro de Estudios Científicos. *Kluwer Academic/Plenum Publishers, New York*, **35**, 101–115.

Rivera, A., Casassa, G., and Acuña, C., 2001. Mediciones de espesor en glaciares de Chile centrosur. *Revista Investigaciones Geográficas*, **35**, 67–100.

Schneider, C. and Gies, D., 2004. Effects of El Niño-southern oscillation on southernmost South America precipitation at 53 degrees S revealed from NCEP-NCAR reanalyses and weather station data. *International Journal of Climatology*, **24(9)**, 1057–1076.

Schrott, L., 1994. *Die Solarstrahlung als steuernder Faktor im Geosystem der Subtropischen semiariden Hochanden (Agua Negra, San Juan, Argentina)*. Ph.D. thesis, Universität Heidelberg.

Thompson, L., Mosley-Thompson, E., Davis, M., LIN, P., Henderson, K., Coledai, J., Bolzan, J., and LIU, K., 1995. Late Glacial Stage and Holocene tropical ice core records from Huascarán, Peru. *Science*, **269(5220)**, 46–50.

Thompson, L. G., Mosley Thompson, E., Bolzan, J. F., and Koci, B. R., 1985. A 1500-Year Record of Tropical Precipitation in Ice Cores from Quelccaya Ice Cap, Peru. *Science*, **229**, 971–973.

Veit, H., 2000. Klima- und Landschaftswandel in der Atacama. *Geographische Rundschau*, **52(9)**, 4–9.

Vose, R., Schmoyer, R., Steurer, P., Peterson, T., Heim, R., Karl, T., and Eischeid, J., 1992. *The Global Historical Climatology Network: long-term monthly temperature, precipitation, sea level pressure, and station pressure data. ORNL/CDIAC-53, NDP-041.* Carbon Dioxide Information Analysis Center, Oak Ridge National Laboratory, Oak Ridge, Tennessee.

Vuille, M., Bradley, R., Healy, R., Werner, M., Hardy, D., Thompson, L., and Keimig, F., 2003a. Modeling $\delta^{18}O$ in precipitation over the tropical Americas: 2. Simulation of the stable isotope signal in Andean ice cores. *Journal of Geophysical Research - Atmospheres*, **108(D6)**.

Vuille, M., Bradley, R., Werner, M., Healy, R., and Keiming, F., 2003b. Modeling $\delta^{18}O$ in precipitation over the tropical Americas: 1. Interannual variability and climatic controls. *Journal of Geophysical Research - Atmospheres*, **108(D6)**, 4174.

Chapter 4

Millennial variability of European climate inferred from an Alpine ice core

ABSTRACT. This study presents the results of the investigation of an ice core from Colle Gnifetti, Monte Rose massif, Swiss Alps (4450 m a.s.l., 45°56′N, 7°53′E). In September 2003, the core was drilled to bedrock, where the ice was expected to be several thousand years old. The last Millennium is highly resolved. An ice core based isotope record is therefore available and can be used for inter comparison with other millennial climate records based on different proxies.

4.1 Introduction

In order to understand natural climate variability on a global as well as on a regional scale there is an urgent need for climate proxies extending our knowledge about climatic variability of the past well beyond the time frame instrumental meteorological data is available for (since around 1850). The information of many such proxies can then be synthesized creating a multi-proxy reconstruction, which is nowadays being regarded state of the art and considered to be much more reliable than interpretations based on a single parameter alone. These reconstructions are not only urgently needed because they help understanding natural climate variability but also because they provide evidence against which regional climate models can be tested. Only by optimizing such models will it be possible to make predictions about e.g. changes in precipitation and temperature on a regional scale. Despite the predictability of climate on a global scale, where sophisticated climate models (AOGCMs) "simulate atmospheric general circulation features well in general" (IPCC - The Scientific Basis, 2001, p. 53), climate models on a regional scale give less consistent results.

From an European perspective such multi-proxy reconstructions are already available (e.g. Luterbacher et al. (2004): 1500–2003 A.D., Glueck and Stockton (2001): 1429–1983 A.D., Moberg et al. (2005): ∼0–1990 A.D.). Many climate reconstructions (e.g. Esper et al., 2002) rely on the temperature-dependent growth rate of trees at high elevation, which is archived in their annual ring width. This archive provides indeed one of the best temperature records, although tree rings generally only record growing season temperature. However, the tree's growth rate also depends on the age of the individual plant (young trees grow faster than old trees) and this has to be corrected for. The transfer from tree ring width into temperature implies the use of sophisticated statistical methods and the way these these methods are applied and calibrated influences the amplitude of the results. In other words, differences in the statistical procedure result in a different magnitude of past change (J. Esper, personal communication and Esper et al. (2004)) whereas high-frequency (short-term) variability is less affected. It would therefore be desirable, to have an independent proxy of similar or superior quality to tree rings allowing for comparison. An isotope record from a European ice core would in principle fulfill these requirements.

The Alps are Central Europe's main mountain range. Their west to east (south to north) extension is about 800 km (150 km) (5.5–16°E, 44–48°N). The highest summit, Mont Blanc, has an altitude of 4807 meters. Although a total of 5422 glaciers exist covering an area of 3,010 km^2 (Paul et al., 2004), only very few of them are suitable for paleo climate research and no long-term record has been available so far. The accumulation zone of most of the Alpine glaciers is found relatively close to the snow equilibrium line, which means that these glaciers experience strong surface melting during summer. The resulting melt water percolation destroys the chemical and the isotopic signature rendering them unsuitable for paleo climate research. Generally, glaciers with sufficiently cold firn temperatures and no or little melt-water percolation can only be found above 4,000 m a.s.l. in the Northern part and above 4,300 m a.s.l. in the Southern part of the Alps (Suter et al., 2001). With this limitation, only the Mont Blanc region, the Monte Rosa Massif and a few sites in the Bernese Oberland are potentially interesting. Out of the few glaciers at such high elevation, many of them are also unsuitable because of their steep surface topography being unfavorable (the ice flow on steep glaciers is too quick and unpredictable) or they experience very high accumulation rates allowing for a good time resolution but limiting the covered time span in many cases to less than 100 years, e.g. on Grenzgletscher (not reaching bedrock, Eichler et al. (2000)) or Col du Dôme, near Mont Blanc, Legrand et al. (2002)). Figure 4.1 shows the location of all these drilling sites.

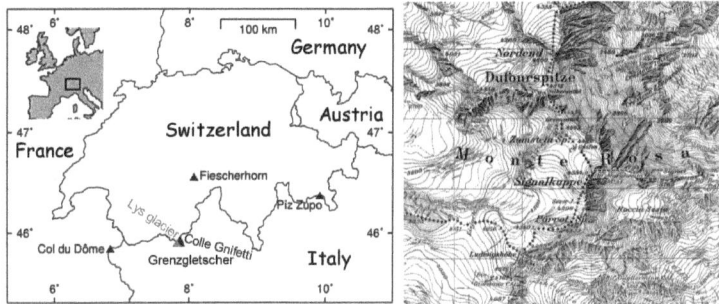

Figure 4.1: Location of the ice core drilling site *Colle Gnifetti* and other ice coring sites in the Alps. The red flag marks the drilling sites of the ice cores drilled on the glacier saddle in September 2003.

The Alps are a very promising study site for ice core research as a dense network of observational data is available (e.g. temperature, precipitation, stable isotopes). This allows for the calibration of obtained records making their interpretation more reliable. Some of the instrumental temperature data reach over 250 years back in time (e.g. Böhm et al., 2001).

4.1.1 The history of ice core drilling on Colle Gnifetti

The potential of the glacier on Colle Gnifetti as climate archive was recognized early. In 1976 and 1977, less than a decade after Dansgaard et al. had published results from the first important Greenland ice core, Camp Century (1969), cores were drilled on Colle Gnifetti (Schotterer et al., 1978).

In 1982, in a project organized by H. Oeschger and U. Schotterer, a total of four cores was drilled there (two drilling sites, two parallel cores from each site). The so called *Red Core* (B82-1) reached bedrock at 124 m, the parallel *Blue Core*, drilled at the same spot, was 109 m long, not reaching bedrock (B82-1). The two parallel *Chemistry Cores* (B82-2) reached bedrock and were 66 m long. The location of the drilling sites is shown in Figure 4.2.

The three cores to bedrock were originally estimated to cover at least the last 500–1000 years. Accumulation rates were found to vary between 0.20 and 0.35 m water equivalent per year. These low values were ascribed to the influence of snow erosion by strong wind. Additionally, the 2,000 m high south-east face of the Monte Rosa-massif acted as an enormous sink for snow (Schotterer et al., 1985). The accumulation rate for the Blue Core was found to be 0.33 m weq. per year (Döscher et al., 1995). The observed accumulation rates values are also lower by a factor of 7–14 when compared to the estimated annual precipitation, which can be inferred from the Grenzgletscher ice core. This core was drilled in 1994 about 900 m away at a relatively wind-protected site (altitude: 4200 m a.s.l. Eichler et al. (2000)). There, the entire yearly precipitation is preserved and in the order of 2.7 m weq. per year.

Most of the Blue Core was analyzed by A. Döscher in the course of her PhD (Döscher, 1996). The data from the Blue Core also served for comparison with this study. Major ions were analyzed over the entire core length (100.8 m), a lead pollution record (1650–1980 A.D.) was provided (Schwikowski et al., 2004) and a small section was analyzed for $\delta^{18}O$ (1950–1982 A.D.).

Concentrations of the most important acidifying species sulfate and nitrate in the Blue Core had strongly increased from the preindustrial age, 1850–1880 A.D., to the industrial time period 1965–1981 A.D., by a factor of 5.8±0.9 and 2.3±0.3, respectively (Döscher et al., 1995). Chloride concentrations showed no trend. Concentrations of ammonia, the primary gaseous alkaline species in the Atmosphere over Europe, increased by a factor of three, indicating increasing emissions in Europe during the 20th century (Döscher et al., 1996). The lowermost part of the Red Core showed an intriguing yellow layer and strongly elevated ionic concentrations (Huber, 1996). The origin of this layer is still not fully understood. In the Chemistry Core, a strong decrease in $\delta^{18}O$ was observed close to bedrock, indicating that the deepest part of the core might contain traces of Pleistocene ice from the last glacial (Wagenbach, 1992).

In 1995, two more cores were recovered on the assumed flow line of the *Chemistry cores* in a field campaign organized by D. Wagenbach. They reached bedrock and had a length of 62 m (KCH, B95-1) and 100 m (KCS, B95-2), respectively (Dällenbach, 2000). Based on the analysis of $\delta^{18}O$ of atmospheric oxygen trapped in air bubbles in the ice it was concluded that the basal ice on Colle Gnifetti was not of Pleistocene origin (Keck, 2001, p. 97).

Finally, in September 2003, a Swiss-Italian team lead by M. Schwikowski drilled two new cores that were 82 m and 81 m long (core 1 and core 2), respectively, providing samples for this study (Schwikowski et al., 2003). The cores were drilled at two sites were the thickness of ice older than 500 years was highest (>14 m) according to ice flow modelling (see Fig. 4.2, M. Lüthi, unpublished). According to this model, the age of ice at bedrock could be in the order of 2000 years.

Considering the relatively small size of the saddle (800 × 800 metres, Fig. 4.2), the inhomogeneity in ice thickness is high which is attributable to the nearby glacier margins (from North-West to South-East).

The two Chemistry Cores were analyzed for a variety of parameters by Wagenbach et al. at University of Heidelberg, Germany. A couple of students used the cores for their thesis (Schäfer, 1995; Armbruster, 2000; Keck, 2001).

Relatively near to Colle Gnifetti, another ice core was drilled on Colle del Lys (see Figure 4.1). It yielded a fifty-year record of organochlorine pesticides (e.g. DDT) (Villa et al., 2003).

This study essentially presents results from core 1, whereas core 2 has so far almost remained untouched.

4.2 Methods

4.2.1 Drilling campaigns

In September 2003, a field campaign was started attempting the recovery of two ice cores on Colle Gnifetti, Monte Rosa massif, Swiss Alps (see Fig. 4.3). The team consisted of a Swiss group headed by Margit Schwikowski and an Italian group. The drilling site was accessed from Switzerland, starting in the village of Zermatt. After spending three nights at Gornergrat ($\sim$3100 m a.s.l.) the team and the equipment were flown to the glacier saddle by helicopter. The team could stay nearby in Europe's highest mountain hut, Rifugio Margherita, 4554 m a.s.l. Two cores were successfully drilled using the light-weight drilling equipment *FELICS* (Ginot et al., 2001b). Drilling was stopped at 81.9 m and 81.1 m, respectively, when it was thought that bedrock had been reached. According to A. Schwerzmann the exact GPS coordinates of the two drilling sites were 45°55'50.41"N, 7°52'33.50"E, 4455 m a.s.l.; Swiss coordinates: 633847.44 ± 0.02 and 86524.66 ± 0.02, core 1, and 45°55'50.38"N, 7°52'33.37"E, 4455 m a.s.l.; Swiss coordinates: 633844.67 ± 0.02 and 86523.94 ± 0.02, core 2.

The two cores were drilled about 2 m apart from each other (Fig. 4.2, CG2003). Core 1 consisted of 125 individually packed core segments (No. 1–125), whereas 124 segments were recovered and packed for core 2 (No. 126–249). The core segments had a length of 70 cm, on average, some of them were shorter. All the samples were stored in insulated boxes on the glacier.

After coring was completed they were flown down to Zermatt and transported in a frozen condition to the cold storage of AZM in Suhr[1] which served as the central ice storage for the ice coring group at Paul Scherrer Institut. Measurements of the borehole temperature were conducted by A. Schwerzmann (VAW, ETH Zürich). For core 1, the sum over the length of

---

[1]AZM, Aargauer Zentralmolkerei, AG, Obertelweg 2, Postfach, CH-5034 Suhr

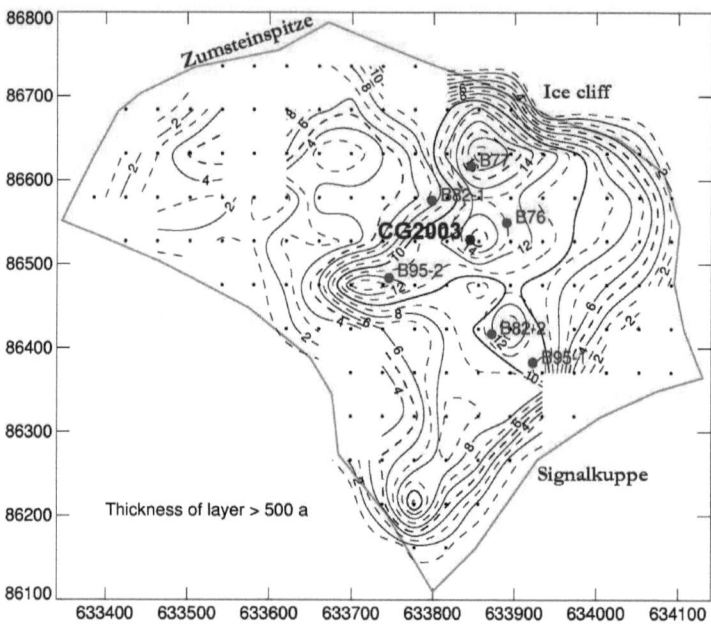

Figure 4.2: Map of Colle Gnifetti showing the estimated layer thickness of ice older than 500 years and the position of all the drilling sites there. B82-1 and B82-2 indicate the drilling sites of the Red Core, the Blue Core and the Chemistry Cores, respectively. B95-1 and B95-2 mark the site of the KCH core and the KCS core, respectively and the blue circle shows the drilling site of the parallel cores drilled in 2003. Core 1 was drilled close to the saddle's point of lowest surface elevation. The dimension of the map is 800 × 800 meters. The figure is adapted from M. Lüthi (VAW, ETH Zürich, unpublished).

Figure 4.3: The glacier saddle Colle Gnifetti as seen from Liskamm. The line points to the drilling site (Core 1). The summit's names are (from left to right): Nordend, Dufourspitze, Zumsteinspitze, Signalkuppe and Parrotspitze. Photographer: Kai Hassler, ©2005.

the individual segments was 80.2 m and thus 1.7 m less than the depth of the bore hole, probably because some ice burst into pieces when drilled and could not be recovered as entire segments. Some uncertainty is also attributable to the depth recording which was done by estimating the length of the cable, the drilling device was attached to.

This sum over the length of the individual segments will be used instead of the physical borehole depth and referred to as length of the core from now on. Practically all the analysis in this study were conducted on core 1, the only exception being samples analyzed for lead-210, those were taken from core 2.

4.2.2 Chemical analysis

The core was processed at -20°C in the cold room at *Paul Scherrer Institut*. A detailed visual stratigraphy was recorded over the entire length of the core recording ice lenses, dust layers and firn structure. All the core segments were weighed and photographed. The density of each segment was calculated by dividing the volume by the mass, assuming a cylinder-shape for the volume calculation. The error of that method is due to limited precision in volume determination because a perfect cylinder shape was rather the exception as small pieces easily broke off when the segments were ripped off the glacier in the course of their recovery. Moreover, the cylinder's base is often not perpendicular to the height. The obtained density values for individual core sections were used to transform the absolute depth scale into a scale based on the depth in meter water equivalent (m weq.). For many parameters, it is more convenient to look at them on a water equivalent scale as this scale corrects for the thinning caused by firn compaction. The depth in m weq. was calculated by multiplying the length of each individual core segment by its density and by summing up over the obtained lengths.

$$\text{depth(m weq.)} = \frac{\text{depth(m)} \times \text{density}}{\text{density of water } (1.00 \text{ kg/dm}^3)}$$

Core 1, that is 80.2 meters long in real depth therefore yields a depth of 62.5 m weq.

Two samples were prepared for lead-210 analysis (Gäggeler et al., 1983) from the core segments 187 to 189 and 211 to 214 (core 2), respectively. Both of the two samples were taken from slightly below a visible dust layer, with the intention to estimate the age of these dust layers. When transferring their respective depth from the depth scale of core 2 to the core 1 scale their depth would correspond to 26.1–27.9 m weq. and 39.6–41.9 m weq., respectively. The sample size had to be relatively large to get an average value for several years, as the seasonal and the decadal variability in lead deposition is high due to fluctuations in the efficiency of transport to the high-elevation sites (Eichler et al., 2000).

Because a lot of work had already been carried out on the Blue Core (drilled in 1982) covering the last few centuries (1650–1980 A.D., Schwikowski et al. (2004)), the primary emphasis for the 2003 core was its very old section and the last 20 years. Therefore, the core processing started at a depth of 54 m, at an estimated age of ∼1900 A.D. according to glaciological predictions by M. Lüthi (unpublished) and was performed to the bottom of the core (80.2 m). In a second step, the section from 0–20 m was processed which should represent 1977–2003 A.D. This assumption was based on a visible dust layer at 19 m attributed to the Saharan dust event of 1977.

Hence, at a depth of 0–20 meters and 54–80.2 meters the segments were cut with a band saw, such that the outer part was removed and the inner part, that was contamination-free, was sampled at an increasing spatial resolution from 6.0 to 1.5 cm. Individual samples were transferred into pre-cleaned containers and stored at -20°C. The samples were then melted and analyzed for major ions (Na^+, NH_4^+, K^+, Mg^{2+}, Ca^{2+}, F^-, CH_3COO^-, $HCOO^-$, $CH_3SO_3^-$, Cl^-, NO_3^-, SO_4^{2-}) by ion chromatography using standard procedures (Eichler et al., 2000) as well as for stable isotopes ($\delta^{18}O$ and δD). Between a depth of 20–54 meters (10.9–38.4 m weq.), samples were prepared from each segment (about 70 cm) and measured for stable isotopes only, because there was insufficient time to sample the entire core at high resolution. The coarse resolution is obvious in the final record, see Figure 4.37 on page 121.

Tritium (^{3}H) measurements were performed between a depth of 23 and 28.6 m by J. Eikenberg, Paul Scherrer Institut. There, the nuclear weapon horizon of 1963 (detectable by a tritium concentration maximum) was expected. Samples were taken for individual core segments ($\sim$70 cm).

An overview of the state of analysis conducted in this study (ion chromatography, stable isotopes and tritium) is illustrated in Figure 4.4. Between 75.5 and 77.0 m, the core quality was poor. The segments there had broken apart during drilling and only small pieces (<2 cm) could be recovered. This proved to be problematic for sample preparation as the decontamination procedure is only applicable for pieces large enough (>5 cm) to be cut with the band saw. No ion-chromatography data is available and δ^{18}O could only partly be determined.

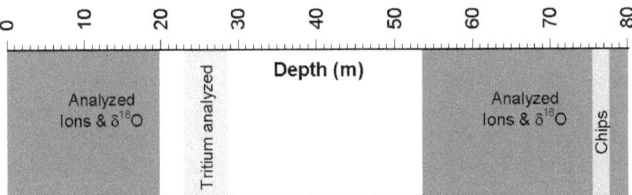

Figure 4.4: Current state of the Analysis on the Colle Gnifetti ice core: *Analyzed* means that the corresponding sections were analyzed at high resolution for major ions and stable isotopes. *Chips* indicates a section of bad ice quality (small pieces). No samples could be prepared for ion chromatography from that section.

For dating the deeper parts of the Colle Gnifetti ice core radiocarbon (^{14}C) results were used. These results were obtained by T. Jenk and are presented in his PhD thesis (in preparation). Details of the method are presented in Jenk et al., submitted in 2005. For this study, carbonaceous particles filtered of the melted samples ($\sim$1 kg of ice each) from several core sections between 52.5 and 62.5 m weq. were used.

4.3 Results and Discussion

4.3.1 Bore hole temperature

Thermistor cable measurements of the bore hole temperature by the VAW, ETH Zürich showed that the temperature of snow and ice in the 2003 bore hole varied between -12°C and -14°C (Fig. 4.5a). These temperatures were the lowest found in a study (Suter et al., 2001) comparing the high-elevation sites of Jungfraufirn (3,400–3,600 m a.s.l., Bernese Alps) and Breithornplateau (3,800 m a.s.l., Valais Alps) and Grenzgletscher (3,900–4,450 m a.s.l., Valais Alps). The firn is therefore cold enough to prevent melt water from percolating through more than a few cm before refreezing. Very few ice lenses were observed. They form when penetrating water refreezes and thus they would be a sign of strong melting. The temperature increase towards bedrock is attributed to geothermal heat flux slowly warming the glacier from below. The effect of strong warming during the 1980s and 1990s (see fig 1.1 on page 12) is not observed, which is not understood. The temperature profile starts at a depth of 10 m corresponding to the year $\sim$1990, as this is the depth heat can penetrate into. Figure 4.5b also shows the temperature profile of bore holes drilled on the saddle in 1982 and 1995, respectively. They all reached bedrock at different positions on the glacier and thus at a different depth. The 1995 profile is practically identical to the profile of the 2003 hole. The hole drilled in 1982 shows a similar pattern but temperatures are all lower by about 0.4°C. This cannot be attributed to warmer air temperatures during the last 20 years because if that was the reason for the difference, then the effect would be dependent on depth and be stronger closer to the surface. Therefore, the temperature difference indicates that the temperature of the glacier varies spatially. This was also demonstrated by Suter (2002, p. 92) who showed that on Colle Gnifetti firn temperatures at 18 m depth increased from -14°C on the northern slope of Signalkuppe to -10°C on the southernly exposed slope of Zumsteinspitze.

4.3.2 Density profile

Due to compaction, the density of the firn (old snow) increases with depth. Because of its high air content, fresh snow has a density of less than 0.2 g/cm^3 but soon after its deposition it starts to compact by recrystallization and by reducing the volume of the pores in the snow under its own weight. The air in the pores can still circulate and thus exchange with the atmosphere. At a density of about 0.83 g/cm^3 (Paterson, 1994, p. 12) the pores are closed off and form air bubbles, which then interrupts the circulation. This critical

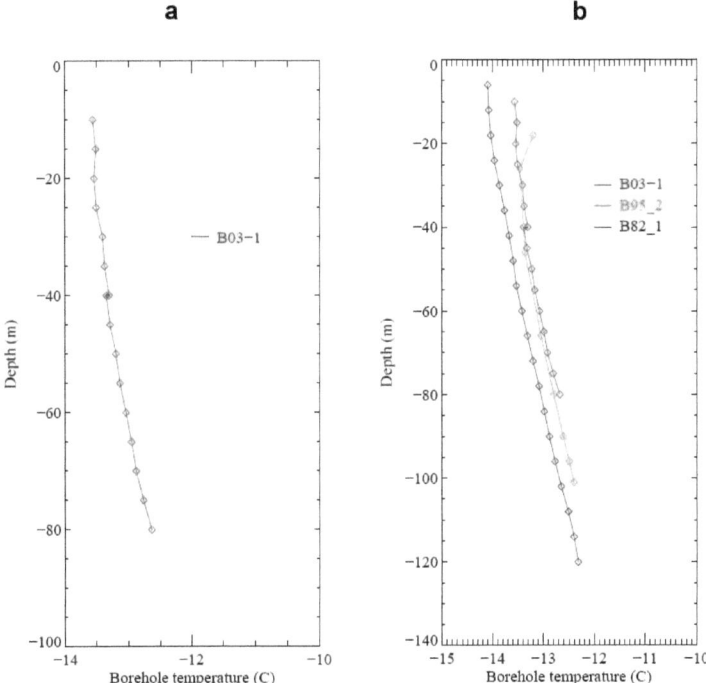

Figure 4.5: Borehole temperature for the Colle Gnifetti 2003 ice core (a) and comparison to bedrock boreholes of different length, drilled earlier nearby (b). The data is adapted from Böhlert (2005).

value also marks the firn-ice transition, although this is not a very sharp boundary.

In the Colle Gnifetti ice core the density increases with depth as expected for a typical glacier (Fig. 4.6). From a depth of 45 meters onwards it levels off at around 0.9 g/cm³, the maximum density for glacial ice. Deviations from this value are due to the limited precision when the segment's volume is calculated. The firn-ice transition takes place at a depth of roughly 35 meters. There is a local density maximum centered at 1.5 meters. The

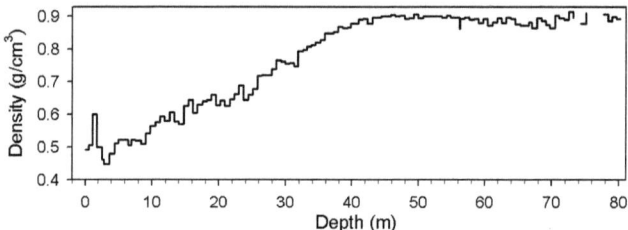

Figure 4.6: Profile of density increase of firn and ice shown for the 2003 ice core from Colle Gnifetti. No values are given for segments of bad core quality (gaps).

high values there can be attributed to percolation and refreezing of melt water as small ice lenses were observed in the corresponding core segment. The multiple occurrence of refrozen water here is probably attributable to the summer 2003 heat wave when June, July and August temperatures in Switzerland were some 5°C warmer than on average, corresponding to more than a 5σ deviation from the 1961–1990 mean (Schär et al., 2004). Colle Gnifetti lies well within the center of the observed anomalies located over the Western Alps.

4.3.3 Dating by annual layer counting, tritium, dust layer stratigraphy, lead-210 (^{210}Pb) and radiocarbon

Annual layer counting

All ions show a seasonal cycle in concentration. This is due to enhanced vertical mixing of atmospheric layers during warmer seasons (e.g. Gäggeler et al.,

1995; Döscher et al., 1996). Consequently, air masses containing aerosol particles are lifted up to high mountains much more efficiently in summer than in winter when atmospheric layers are poorly mixed. Sources for these aerosols are in most cases located at low elevations where industrialized centers are encountered. The seasonality in ion concentration is most pronounced for species whose production is itself varying with the seasons (e.g. for ammonium). Seasonally varying parameters can be used for determination of annual layers in the ice core. As the use of a single parameter for the identification of such layers is sometimes ambiguous it is recommendable to use a multi-proxy approach. In this case the identification of annual layers is based on the combined interpretation of the ammonium concentration and $\delta^{18}O$ which is illustrated in Figure 4.7. Years were identified where maxima in both signals occurred simultaneously based on the idea that the strongest NH_4^+ input and highest temperatures occur in summer. In some cases such a coincidence was missing and years were ascribed to the more pronounced maximum in either one of the two signals. E.g. between 4 and 5 m weq., maxima are missing in $\delta^{18}O$ but they are also not strongly pronounced in ammonium which was used there for the layer identification. Annual layer

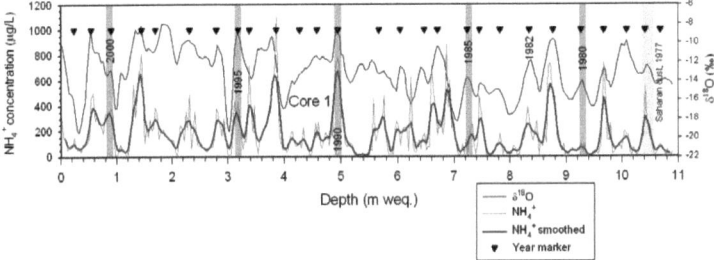

Figure 4.7: Annual Layer counting based on seasonal variations of $\delta^{18}O$ and ammonium concentration. Year markers were set on maxima occurring in both signals where such a coincidence could be determined or they were set on the more pronounced maxima of a single species. Grey bars indicate intervals of 5 years, the yellow bar indicates the 1977 Saharan dust input.

counting is only possible when high resolution data (10–15 samples per year) is available. Therefore it is limited to the 1976–2003 section of the core. At depths greater than 38.4 m weq., annual layers are only about 8 cm weq. thick and, as the sampling resolution there was 1.9–2.3 cm weq., sub-annual resolution was practically lost.

Annual layer counting was also successfully performed for the Blue Core for the time period 1963–1977 A.D. (Döscher et al., 1996). There, it was based only on ammonium concentration.

Tritium

A very good reference horizon that was identified was the maximum in tritium activity concentration at a depth of 14.2 meters weq. (see Fig. 4.8). This maximum is due to intense nuclear weapon testing in 1962 . That year, the Russian army blew up a large amount of hydrogen bombs before a Soviet-American treaty came into force in 1963 banning tests in outer space, in the atmosphere and under water (detonation underground was still allowed). On the northern hemisphere, the maximum tritium concentration in precipitation was reported in 1963. In 1963, when the tritium-enriched snow was deposited, concentrations were roughly ten times higher than at the time of the measurements in January, 2005 (the half life is 12.3 years). The decay-corrected peak value for the 2003 core is 1390 TU, less than the peak value observed in the Blue Core (1920 TU, Döscher (1996)). However, the sampling resolution was much higher for the Blue Core which accounts for the higher concentration: When averaged over 70 cm, corresponding to the sampling resolution of the 2003 core, the concentration in the Blue Core would only be 1090 TU.

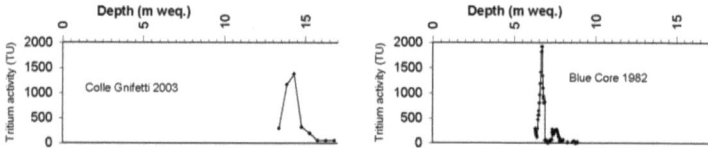

Figure 4.8: Tritium activity in the 2003 core and in the Blue Core. The values are decay-corrected.

In the 2003 core, the width of the peak is ∼1.4 m weq., whereas it is only 0.6 m weq. in the Blue Core. At the drilling site of the Blue Core the accumulation may have been lower in 1963 or the peak was not fully sampled

Dust layer stratigraphy and lead-210

In sporadic events, dust from the Sahara desert in North-Africa is deposited over central Europe. This happens, when the dust is lifted up during strong desert storms and the particles are transported in the mid-troposphere on trajectories to Europe. The particles form yellow layers when they are deposited on a glacier. Some of the strong events were historically documented, e.g. the dust input of 1936 and 1901/1902 (Götz, 1936). For these two events, the typical particle size ranged from 1 to 100 μm and the quantitative distribution maximum was around 5 μm. Main mineralogical compounds of the dust were carbonates, muscovite, biotite and quartz. Studies by Wagenbach and Geis (1989) confirmed this and found that the maximum of the particle size distribution for 16 investigated Saharan dust layers varied between 2.5 and 10 μm.

Some yellow layers that were observed in the Colle Gnifetti 2003 ice core could be used as reference horizons for dating by comparing them to the data from the Blue core where dating was corroborated by lead-210 analysis and the identification of volcanic horizons (Döscher, 1996). Figure 4.9 shows the dust horizon from the observed event in the year 1977 (described by Prodi and Fea (1978)). Such layers are easily visible by eye, although it is more difficult to recognize them on a photograph.

Figure 4.9: Picture of core segment 33 of the Colle Gnifetti 2003 ice core. The yellow layer formed during Saharan dust deposition in 1977 and is clearly visible between 3 and 15 cm. In order to intensify the contrast, colors were enhanced and the brightness was increased.

An attempt was made to visually match observed dust layers in the 2003 core to dust layers found in the Blue Core. As no visual stratigraphy existed for the Blue Core, yellow layers and sections of high calcium concentration in the new core were attributed to high values of calcium (Ca^{2+}) ion concentrations in the 1982 core which is shown in Figure 4.10, page 87. Ca^{2+} is an excellent tracer for mineral dust. The problem associated to matching Ca^{2+} concentration in one core to visually observable layers in another core lies in the fact that high Ca^{2+} concentrations are not always visible in stratigraphy in the form of yellow layers.

The detailed calcium profile (not shown) of the 1977 event shows a double peak in the Blue core (Döscher, 1996) and the dust event at 10.5 m weq. in the 2003 core also shows a double peak. Because of this and the proximity to the tritium horizon this event can be well assigned to the event of 1977. From 10.9–38.4 m weq., there no is chemistry data yet. Thus, the suggested assignation of visible layers in the 2003 core to high calcium concentration in the Blue core is in somewhat speculative.

For the section 38.4–62.5 m weq. the calcium peaks in the 2003 core were assigned to calcium peaks of the Blue Core such that similar patterns in both cores were matched. The peak at 39.2 m weq. (2003 core) is followed by a sequence of three consecutive peaks centered at 40.8 m weq. and a similar feature is observed in the Blue with a peak at 51.7 m weq. followed by three consecutive peaks centered at 55.2 m weq. The peak at 44.9 m weq. (2003 core) was assigned to the peak at 62.5 m weq. (Blue Core). The three consecutive peaks centered at 48.6 m weq. (2003 core) were assigned to the three consecutive peaks centered at 70.6 m weq. in the Blue core and the peak at 50.9 m weq. in the 2003 core was assigned to the peak at 76.5 m weq. in the Blue Core. Table 4.1 shows the results of the assignation of the two calcium concentration profiles with the respective years derived from the dating of Döscher (1996), 1755–1981 A.D., refined by annual layer counting by M. Schwikowski up to the year 1470 A.D. Dust layer matching based on visual stratigraphy only (no calcium concentration available, 10.9–38.4 m weq.) was not included for dating.

Close to bedrock of the 2003 core an intense yellow layer was found, described in further detail in Section 4.3.12.

Table 4.1: Matching of Ca^{2+} peaks observed in the 2003 core to Ca^{2+} peaks in the well-dated Blue Core (dating uncertainty ± 20 years before 1780 A.D.). Depths are in m weq.

Depth (Blue Core)	Year (Blue Core)	Depth (2003 core)
2.0	1977	10.5
51.7	1764	39.2
55.2	1735	40.8
62.5	1658	44.9
70.6	1572	48.6
76.5	1502	50.9

Based on the Saharan dust event of 1977 and the nuclear weapon horizon,

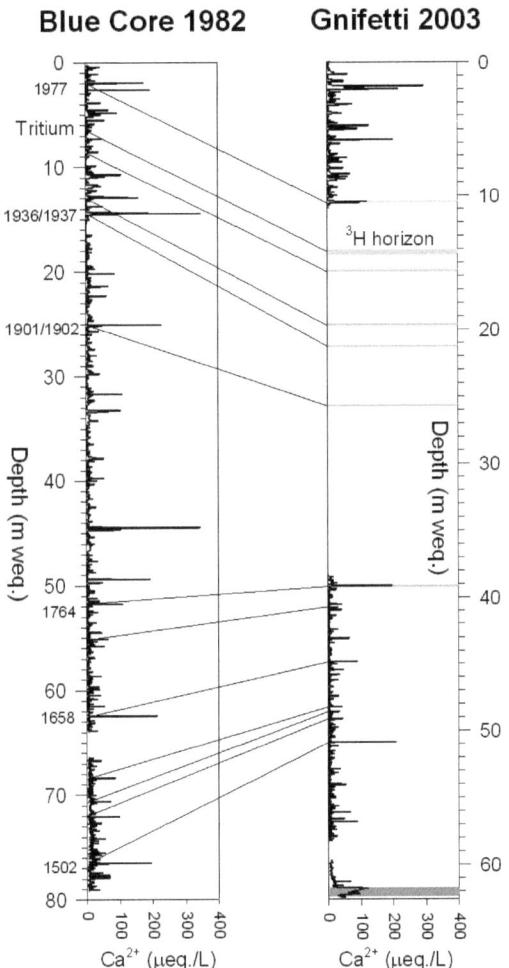

Figure 4.10: Matching of Saharan Dust events between the well-dated Blue Core (drilled in 1982, left) and the 2003 core (right). Dust layers were directly observed by visual stratigraphy (indicated by bars) and partly by Ca^{2+} concentration in the new core, whereas Ca^{2+} concentration was taken as a dust tracer in the Blue Core. The dating uncertainty in the Blue Core is ±20 years for the section older than 1780 A.D. (Schwikowski, personal communication). The tritium horizon, was included here, too.

1963, the average accumulation rate would be 0.40 m weq. (1977–2003) or 0.36 m weq. (1963–2003), respectively. Both values are potentially an underestimation of the "true value", as thinning due to vertical strain might already be active. A slightly lower accumulation rate of 0.33 m weq. was found for the Blue Core (Döscher et al., 1995).

According to the matching, the dust layers at a depth of 25.7 and 39.2 m weq. were ascribed to the events of 1901/1902 and 1764, respectively. In order to independently check this, samples slightly deeper than the horizons were prepared from segments 187–189 and 211–214 (Core 2), respectively, and analyzed for lead-210 (^{210}Pb). The results are shown in Table 4.2 and compared to values from the Blue Core (Table 4.3, Döscher (1996, p. 32)). The relative depths of these samples were converted to the Core 1 depth scale.

Table 4.2: ^{210}Pb activity of two core sections located below visible dust horizons at 25.7 and 39.2 m weq., respectively.

Sample	^{210}Pb activity $\frac{mBq}{kg\ ice}$
Sample 1 (26.1 – 27.9 m weq.)	4.7±0.3
Sample 2 (39.6 – 41.9 m weq.)	2.0±0.2
Blank	2–3

Table 4.3: ^{210}Pb activity at specific horizons in the 1982 Blue Core. The data is adapted from Döscher (1996).

Depth (m weq.)	Year	^{210}Pb activity $\frac{mBq}{kg\ ice}$
2.0	1977	42
14.5	1936	14.5
25.1	1901/1902	4.7
>35	older than ~1850	2.0

The assignation of the 1901 and 1764 dust events observed in the Blue Core to the two dust layers at 25.7 and 39.2 m weq. in the 2003 core is therefore consistent with the lead-210 activity there. In principle, a lower activity would be expected for similar horizons in the 2003 core when compared to the 1982 core, because 9–10 years had elapsed between the ^{210}Pb analysis on the 2003 core and on the 1982 core. ^{210}Pb has a half life of 22.3 years, such that after 9.5 years the activity should have decreased by 26%. However, the agreement is still considered good when the variability of lead-210 activity is taken into account. An activity of 2.0 mBq corresponds to the activity measured in blanks such that the age estimation for these samples is limited to "older than 1850".

Dating by volcanic horizons

The use of volcanic horizons was an important part in dating the Blue Core. There, the sulfate/calcium ratio was used to identify volcanic eruptions (Döscher, 1996) of 1912, 1883, 1815 and 1783.

Figure 4.11 shows the sulfate/calcium ratio for the 2003 core for the preindustrial section. Peaks centered at 42.7, 44.8 and 47.9 m weq. showed a ratio higher than 7 but for all the peaks the absolute sulfate concentration was lower than 125 µg/kg, less than what is expected for volcanic horizons: In the Blue Core the sulfate concentrations at the volcanic horizons ranged between 380 to 1170 µg/kg. Before 1783 A.D. no volcanic eruptions were

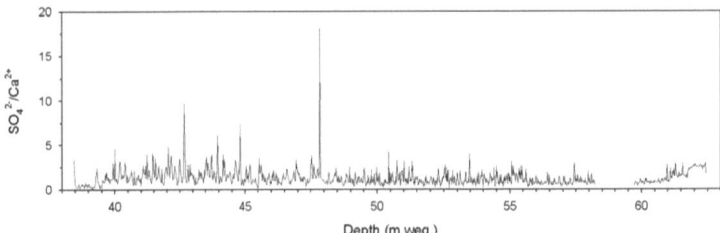

Figure 4.11: Sulfate/calcium ratio for the 2003 core, preindustrial section.

identified in the Blue Core. This is another indication that the section 38.4–62.5 m weq. is probably older than 1783 A.D. which would explain why no volcanic horizons were found in the 2003 core although some were found in the Blue Core.

Radiocarbon

Radiocarbon ages used for dating were taken from the PhD thesis of T. Jenk (in preparation). The sample from the lowermost core segment 125 was taken from the section containing the intense yellow dust layer described in Section 4.3.12, leaving out the lower part of the segment consisting of clear ice. The results for organic carbon (OC) were used and are shown in Table 4.4. The lower and upper limit represent the 1σ-range of the calibrated radiocarbon age.

Table 4.4: Calibrated radiocarbon ages (organic carbon, OC) for the samples from Colle Gnifetti. The depth is in m weq. Radiocarbon dates are calibrated ages before present (before 1950). These dates were provided by T. Jenk. The age for the lowermost core segment 125 is a conservative lower limit.

Segment	Depth	Calib. Age	Limits (1σ)
114	56.46	1100	700, 1400
117	58.07	1200	900, 1550
121	59.86	1750	1550, 1950
123	61.08	3400	3240, 3570
124	61.64	7950	7350, 8550
125	62.18	>11700	2σ

4.3.4 Dating by flow modelling using the stratigraphic horizons and the radiocarbon ages

Combining the results from dust stratigraphy, tritium measurements and radiocarbon measurements, several time horizons were obtained and could be used for dating the 2003 core (Table 4.5 and Figure 4.12).

Age-depth model

In order to establish a continuous age-depth relationship for a core, ice flow models can be applied. Here, a two parameter flow model (originally suggested by Bolzan (1985), refined by Thompson et al. (1990)) proved to match the horizons well. The model is based on a simple analytical expression for the decrease of the annual layer thickness $L_{(z)}$ with depth :

$$L_{(z)} = b\left(1 - \frac{z}{H}\right)^{p+1}$$

z: depth (m weq.), H: glacier thickness (m weq.), b: accumulation rate (m weq.), p: thinning parameter:

The age ($T_{(z)}$) as a function of depth can be easily calculated when the inverse layer thickness is integrated over depth:

$$T_{(z)} = \int \frac{dz}{L_{(z)}} = \frac{1}{b}\int \left(1 - \frac{z}{H}\right)^{-p-1} dz$$

Table 4.5: Reference horizons used for dating. The depth is given in m weq. and the age in years. Radiocarbon dates are calibrated ages before present (before 1950). These dates were provided by T. Jenk.

Horizon type	Depth	Year	Age	min. Age	max. Age
Top of the core	0	2003 A.D.	0	0	0
Saharan dust	10.50	1977 A.D.	26	26	26
Nuclear weapon	14.22	1963 A.D.	40	40	40
Saharan dust	39.19	1764 A.D.	239	239	239
Saharan dust	40.81	1735 A.D.	268	268	268
Saharan dust	44.86	1658 A.D.	345	345	345
Saharan dust	48.58	1572 A.D.	431	431	431
Saharan dust	50.86	1502 A.D.	501	501	501
^{14}C	56.46	∼850 A.D.	1100	700	1400
^{14}C	58.10	∼750 A.D.	1200	900	1550
^{14}C	59.86	∼200 A.D.	1750	1550	1950
^{14}C	61.08	∼1450 B.C.	3400	2850	3550
^{14}C	61.64	∼6000 B.C.	7950	3160	3450

Solving the integral gives

$$T_{(z)} = \frac{H}{bp}\left(1 - \frac{z}{H}\right)^{-p} + C$$

Using $T_{(0)} = 0$, the final age-depth relation is obtained:

$$T_{(z)} = \frac{H}{bp}\left\{\left(1 - \frac{z}{H}\right)^{-p} - 1\right\}$$

The thinning rate (vertical strain rate), an important glaciological parameter, is the first derivative of the layer thickness:

$$L'_{(z)} = \frac{dL_{(z)}}{dz} = -\frac{b(p+1)}{H}\left(1 - \frac{z}{H}\right)^{p}$$

The model has two degrees of freedom, the accumulation rate and the thinning parameter p.

Application of the model on the reference horizons

The model was fit through all the reference horizons (Table 4.5) by varying the thinning parameter p and the accumulation rate b using a least squares approach. Best fit was obtained with an accumulation rate of $b = 1.26 \pm 0.39$

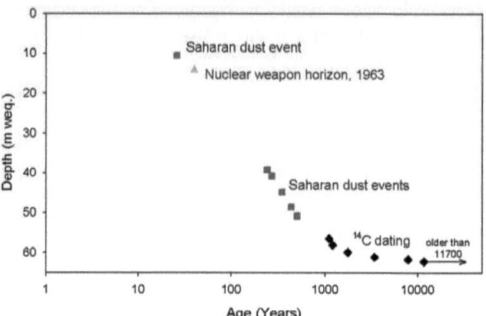

Figure 4.12: Reference horizons obtained by different methods. A logarithmic scale for the age-axis was used. All the horizons but the lowermost were used for dating by flow modelling

m weq./year and a thinning parameter of $p = 1.22 \pm 0.09$. The goodness of fit was $r^2 = 0.984$, the reduced chi-square[2]: $\chi^2_{red} = 92384$. Although r^2 suggested a good fit, the uncertainty of the accumulation rate was relatively high, and the value itself was very different from the value obtained by the reference horizons (0.40 m weq. and 0.36 m weq., respectively). Due to the nature of the model, $T_{(z)}$ is a steep curve at high depths (big age change produced by small depth change) although optically the curve appears flat at high depths (Figure 4.13). This effect is just due to the fact that the dependent variable, the age, is plotted on the x-axis, by convention. Therefore, the distance between the horizons and the model is bigger at greater depths because the least square approach minimizes the vertical distance. This gives larger weight to horizons at greater depth. Moreover, because of the higher absolute numbers and the larger uncertainty related to radiocarbon dates as compared to the other horizons, horizons at great depth have a much stronger influence on the model's parameters. In this case the model basically tries to match the lowermost horizon (7950 years BP) as it has the highest weight.

In order to overcome this effect, the model was calculated using the logarithm of the age values. The resulting parameters for the least square fit were: $b = 0.44 \pm 0.03$, $p = 0.84 \pm 0.04$, $r^2 = 0.994$, $\chi^2_{red} = 0.01902$. The resulting continuous age-depth relationship is shown in Figure 4.13, solid line.

---

[2]Chi-square (χ^2) is the sum of all the squared residues. Reduced chi-square is χ^2 divided by the degrees of freedom, i.e. number of horizons − parameters

Using the logarithmic ages resulted in a higher r² and, more importantly, in a reduced uncertainty of the parameters p and b. The obtained accumulation rate was in agreement with the observations (0.40 and 0.36 m weq./year). The model was now more stable. The value for χ^2 is not comparable as it strongly depends on the absolute numbers used for modelling. The approach of using the logarithmic values of the ages is therefore regarded appropriate.

In order to estimate the dating error, minimum (maximum) values of the parameter b and p were used to constrain minimum and maximum ages. In order to obtain minimum (maximum) ages, the higher (lower) range for b and the lower (higher) range for p was taken (using the standard error for b,±0.03, and p, ±0.04): The model for minimum ages was defined by b=0.47 and p=0.80, whereas the model for maximum ages was defined by b=0.41 and p=0.88. The resulting curves are shown in Figure 4.13, the solid line represents the best fit, the dotted lines represent the constraints for minimum and maximum ages, respectively. The resulting dating error is asymmetric and increases with depth (and age). Table 4.6 shows the dating error at some different depths.

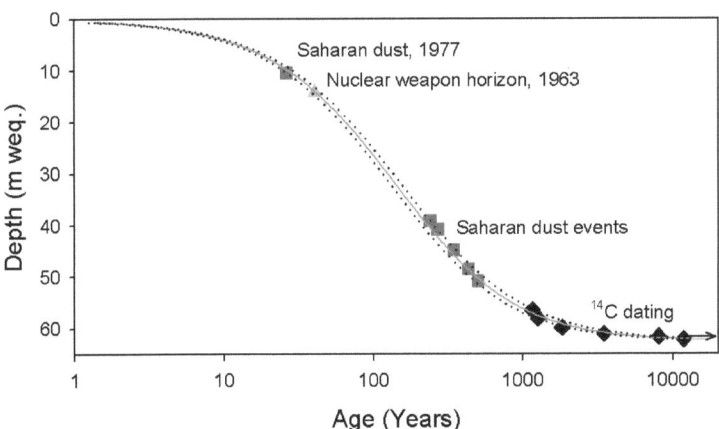

Figure 4.13: Reference horizons and estimation of the age by a 2-parameter flow model. The dotted curves give an estimation for the upper and lower limits of the age, respectively.

The two parameter model assumes non-linear thinning: The decrease of the annual layer thickness is a non-linear function of depth as is the thin-

Table 4.6: Dating error at different depths (m weq.) of the core. The values were rounded.

Depth	Age	−	+
10	26.8	1.8	2.1
40	230	20	23
50	488	49	58
55	850	95	115
60	2380	340	420

ning rate (vertical strain rate). This non-linearity allows for modelling the existence of a considerable amount of old ice near bedrock. The widely used Nye model (Nye, 1963), in contrast, assumes a linear decrease of the annual layer thickness or a constant thinning rate and the accumulation rate is the model's only degree of freedom (here again the relation between layer thickness and age is $T_{(z)} = \int \frac{dz}{L_{(z)}}$):

$$L_{(z)} = \frac{b(H-z)}{H} \qquad T_{(z)} = \frac{H}{b} \times \ln\frac{H}{H-z}$$

The Nye modell cannot account for large amounts of old ice: Assuming that the accumulation rate corresponds to 0.7% of the glacier thickness, as is the case for the investigated core, ice older than 650 years, would only be found in the lowermost 62 cm weq. of the core (1% of the core length). This is incompatible with the age derived from radiocarbon (Table 4.4): Based on the ^{14}C horizon at 58.1 m weq. the lowermost 3.4 m weq. of the core are older than ∼1200 years. Therefore, the 2-parameter model was used (see Fig. 4.13), as this model can account for any amount of old ice.

Another model, suggested by Dansgaard and Johnsen (1969) was also tested. This model assumes a constant strain rate up to a certain depth h, then the strain rate decreases linearly. This parameter was varied using a least square approach. It was found to be 21 m weq. Figure 4.14 shows the corresponding age-depth relationship. The model matches the horizons less well, it overestimates the age between 0 and 15 m weq. and underestimates the age between 39 and 56 m weq. Therefore, the 2-parameter model was used.

Because of the enormous dynamic time range (0–20,000 years, see Fig. 4.13), a logarithmic scale was used for the plot. For comparison, Figure 4.15 shows the same data on a linear age scale covering 4,000 years. The convincing match between data and model allows for the use of the model as transfer function yielding an age for each depth.

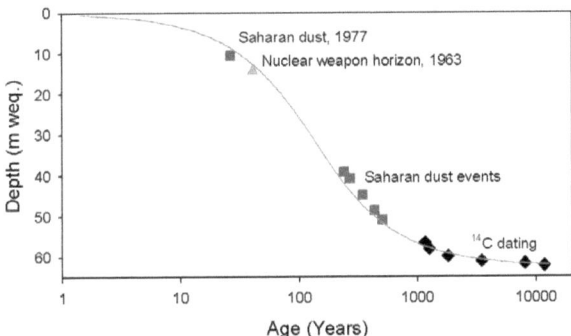

Figure 4.14: Reference horizons and estimation of the age by a flow model suggested by Dansgaard and Johnsen (1969).

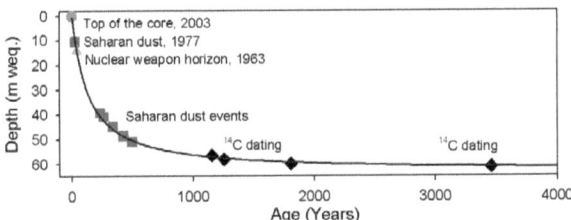

Figure 4.15: Age - depth relation using a linear time axis for the age range 0–4,000 years.

It would be tempting to use these models for the estimation of the age of the ice at bedrock itself. As the ice in cold glaciers is frozen to bedrock and thus immobile, this age should correspond to the date of the formation of the glacier, provided that ice from that time is still preserved (which is often not the case[3]). From a climatological and glaciological standpoint this question is definitely one of the most intriguing ones to be answered for a specific glacier. However, one of the properties of the 2-parameter and the Nye model is that they theoretically predict indefinitely old ice at bedrock This is the result of indefinitely thin layers there:

$$\lim_{z \to H} T_{(z)} = \infty$$

Therefore, these models cannot answer the question of when the glacier was formed.

4.3.5 Stable Isotopes

Figure 4.16 shows the isotope record ($\delta^{18}O$) over the entire core length. The mean $\delta^{18}O$ value for the time period 1976–2003 is -13.1±2.4. For all the samples it would be -13.9±2.1‰(±1σ), also shown in the figure (4.16). The uppermost meters (0–6 m weq.) lie well above the average as well as the section 50–58.3 m weq. From 60.2–62.5 m weq., there is a section of a remarkably depleted signal, which is discussed later.

The average value of -13.9‰ is relatively enriched (high) considering the elevation of the site (4450 m a.s.l.). At the "nearby" meteorological hydrological station at Grimsel pass (46°34′12″N, 8°19′48″E, 1950 m a.s.l, 79 km from Colle Gnifetti), the mean value is -13.7±3.4‰ for the time period 1970–2004 (monthly values; unweighted). The values are therefore practically identical despite a difference in altitude of 2500 meters.

Different altitudinal laps rates for $\delta^{18}O$ found in the literature range from 1.0 to 2.5‰ per 1000 m (Clark and Fritz, 1997, p. 71). Two values from study sites not too far off were 2.0‰ per 1000 m (Jura Mountains, Siegenthaler et al. (1983)) and 1.9‰ per 1000 m (Schwarzwald, Dubois and Flück (1984); cited in Clark and Fritz (1997)). An investigation of the altitudinal influence on $\delta^{18}O$ for the Swiss Alps at sites with elevations ranging from 700 to 4,000 m a.s.l. also confirmed a laps rate of 2.0‰ per 1000 m (Schotterer et al., 1997). This value was therefore taken for further calculations: By extrapolation, the mean value at 4450 meters a.s.l. should be between -16.9‰ and -18.7‰ based on the station data at Locarno (46°10′12″N 8°46′48″E,

---

[3]The ice sheet on Greenland formed ~2.7 million years ago, but the oldest ice found there at bedrock was only 123,000 years old (Andersen et al., 2004)

379 m a.s.l., mean $\delta^{18}O = -8.8 \pm 3.7‰$) and Grimsel, respectively. The high values at Colle Gnifetti are a clear hint that the glacier saddle does not accumulate precipitation all year round but mainly during summer, when precipitation is isotopically enriched. The absence of cold-season accumulation is attributable to the erosion of dry winter snow (Schotterer et al., 1985; Schotterer, 2004). The higher standard deviation in $\delta^{18}O$ at Locarno and Grimsel also indicates a more pronounced seasonal difference because of a stronger seasonality in $\delta^{18}O$ values. For comparison, the mean $\delta^{18}O$ for the Grenzgletscher ice core (4200 m a.s.l.), where accumulation takes place throughout the entire year, is -17.20±4.15 (Eichler et al., 2000).

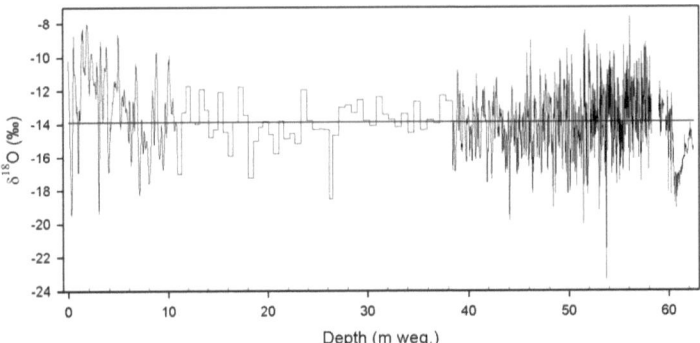

Figure 4.16: $\delta^{18}O$ record over the entire Colle Gnifetti ice core. The vertical line represents the mean of all the samples. The section between 10.9 and 38.4 m weq. was sampled at coarse resolution of ~70 cm. Small sections between 58.3 m weq. and 59.7 m weq. have not yet been analyzed because sample preparation was hindered by the bad core quality.

Figure 4.16 shows the $\delta^{18}O$ record on a depth scale, Figure 4.17 shows the same record on a time scale (550–2003 A.D., corresponding to the section 0–58.25 m weq.). Several features are remarkable when comparing these two figures: Samples were prepared at coarse resolution between 10.9 m weq. and 38.4 m weq., corresponding to the time period 1800–1976. These samples are all about the same "real" size (~70 cm) but because of the increasing density (from 0.63 kg/dm^3 to 0.90 kg/dm^3) the sample size increases from 44 cm weq. to 62 cm weq. Because of the thinning, annual layers are about 31 cm weq. thick near 11 m weq., whereas, near a depth of 38 m weq., their thickness is

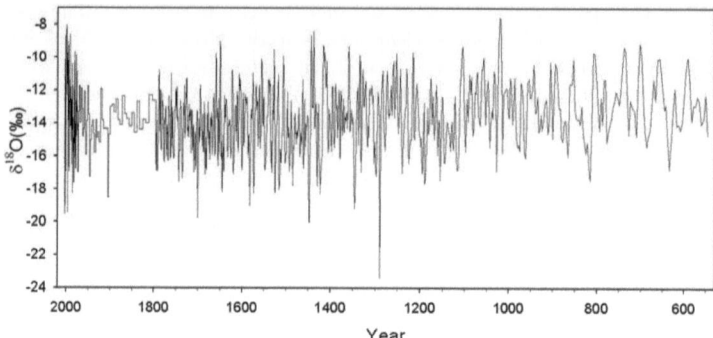

Figure 4.17: Colle Gnifetti δ^{18}O record for the time period 550–2003 A.D. (0–58.25 m weq.). The section of coarse resolution between 10.9 and 38.4 m weq. accounts for less than a sixth of the record length.

only 8 cm weq. As a consequence of both effects, samples comprise more time near 38 m weq. (by a factor of 5–6) and appear thus "bigger". On a depth scale, the section of coarse sampling spans half of the core (Fig. 4.16), in the time domain, however, this section only accounts for 12% of the observations for the time period 550–2003 A.D.

Because of the different sampling resolution and because of the decreasing width of annual layers, the temporal resolution of the data is inhomogeneous. Between 2003 and 1976, the average resolution is 12 samples per year, from 1976 to 1800 it decreases from 0.73 to 0.13 samples per year and between 1800 and 550 A.D. it decreases from 3 to 0.2 samples per year. Overall, there is a general trend towards lower resolution with depth.

Figure 4.18 shows the δ^{18}O record over the last 10,000 years. Here, the depleted signal from 3,000–10,000 years is the dominant feature. The origin of the signal is not yet understood. First, it was thought that such a depletion could be attributed to the presence of ice of the last glacial. This was the interpretation for a layer of depleted stable isotopes values observed on Huascarán, Peru (Thompson et al., 1995), and Sajama, Bolivia (Thompson et al., 1998). According to the current dating such a possibility can be ruled out as the depletion starts by far to early.

An isotopically depleted layer was also observed in the Chemistry Core (Wagenbach, 1992), the KCH core and the KCS core (Keck, 2001). Keck concluded that the lowermost section (89–100 m) of KCS was not of Pleistocene

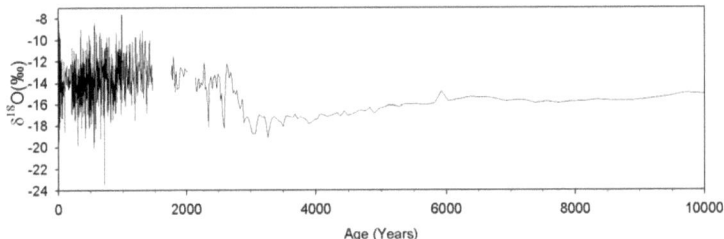

Figure 4.18: Colle Gnifetti $\delta^{18}O$ record of the last 10,000 years.

origin because the $\delta^{18}O$ analyzed on air bubbles trapped in the ice showed values rather typical for the Holocene (-0.4 – 0.4‰) admitting however, that the obtained values "may be subject to artifacts".

He suggested that migration of isotopes as a possible explanation. Liquid water at grain boundaries (especially triple junctions) would be isotopically depleted in equilibrium with water (Lehmann and Siegenthaler, 1991). Keck suggested that the depleted water would move from the zone of simple shear into the zone of pure shear at bedrock (see Figure 4.19). While this theory sounds plausible in principle the driving force is not really explained nor is it discussed why this effect should be relevant on Colle Gnifetti but in other ice cores the interpretation of depleted layers as Pleistocene/Holocene transition would still be valid.

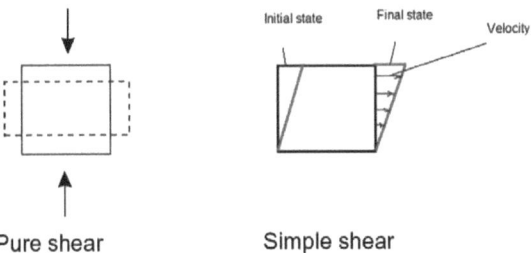

Figure 4.19: Difference between zones of pure shear (pure compression) high above bedrock and simple shear (with horizontal displacement) close to bedrock. The left illustration (pure shear) is from Keck (2001), the right illustration is from Wikipedia (http://en.wikipedia.org/wiki/Simple_shear, March 2006).

Deuterium excess

The deuterium excess, d, is defined by $d = \delta D - 8 \times \delta^{18}O$ (Dansgaard, 1964). It is a measure for non-equilibrium processes during the evaporation process over the moisture source (mainly the ocean). Lower values for d mean that the evaporation process took place under conditions closer to equilibrium conditions. Full equilibrium conditions would require a relative humidity of 100%. Deuterium excess has been interpreted in many different ways, as different factors prevail at different sites. In Antarctica, it was thought that d represents relative humidity at sea level (Jouzel et al., 1982), in Greenland changes in d were related to changes in shifts of the geographical locations of moisture source areas (Masson-Delmotte et al., 2005) and in an ice core from Mongolia high d values were thought to reflect the extreme continentality of the site (Schotterer et al., 1997).

For this study, more than two thirds of the ice core have been analyzed for δD. Figure 4.20 shows the corresponding record for the entire core length. There is a clear correlation between $\delta^{18}O$ and d-excess for the time period 1976–2003. This is obvious from Figure 4.20b. Figure 4.21 shows the corresponding scatter plot. The square of the correlation coefficient, r^2, is 0.30 whereas it is much lower for the preindustrial section of the core (r^2=0.09, samples older than 200 years). This means, that for the time period 1976–2003, the d-excess is generally higher during the warmest part of the accumulation season.

Disregarding three consecutive samples with a value for d between -9 and -11 observed in the year 2002 (at a depth of 1 m weq.), the values are almost symmetrically distributed around the mean. The distribution is slightly leptokurtic (skewness<0.3, kurtosis=1.6, see Figure 4.22), i.e. the peak is steeper than what would be observed for a normal distribution (where kurtosis=0). The average value for all the 1204 samples is 12.9±2.1‰ (1976–2003: d=12.7±2.3‰), which means that the three consecutive samples with a value for d between -9 and -11 are more than 10σ below the mean of the remaining values. This might be due to instrumental problems, however, negative d-excess values are also rarely observed in the Swiss stations, where stable isotopes in precipitation are recorded (see Section 4.3.6). For 1600–2003 A.D. (0–48 m weq.), the d-excess record is almost complete, there is no long-term trend in the data and the mean value for d is 13.0±2.0‰.

The excellent correlation between the two stable isotope parameters $\delta^{18}O$ and δD (r^2=0.99) is shown in Figure 4.23. The slope of the regression line (8.5) is relatively close to the Global Meteoric Water Line (8, Dansgaard (1964)), excluding the three samples with extraordinarily low d-excess.

The value for the average deuterium excess is high when compared to an

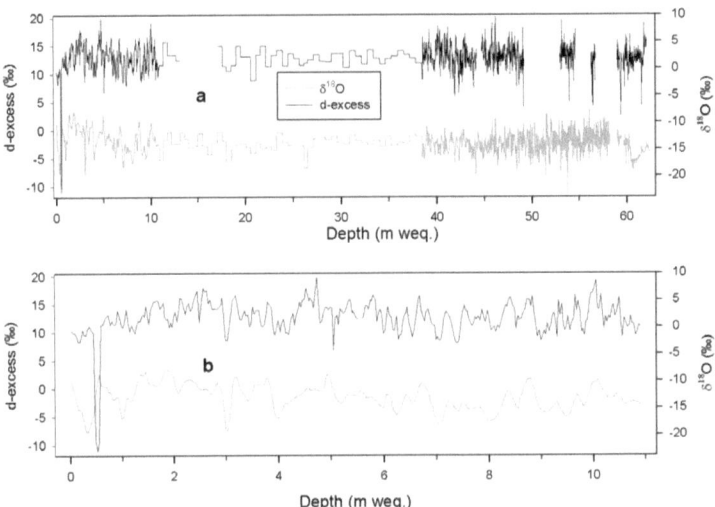

Figure 4.20: Colle Gnifetti record of deuterium excess and $\delta^{18}O$ for (a) the entire core length and (b) the uppermost section (0–11 m weq., 1976–2003 A.D.). The negative values centered at 1 m weq. might be due to instrumental difficulties.

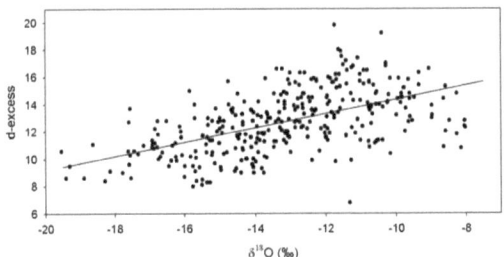

Figure 4.21: Colle Gnifetti deuterium excess vs $\delta^{18}O$ for the time period 1976–2003 A.D.

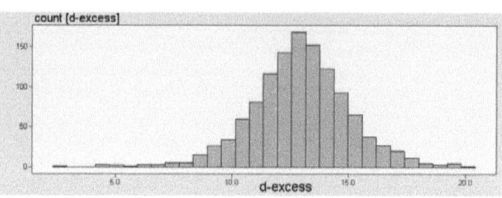

Figure 4.22: Histogram for deuterium excess observed on Colle Gnifetti.

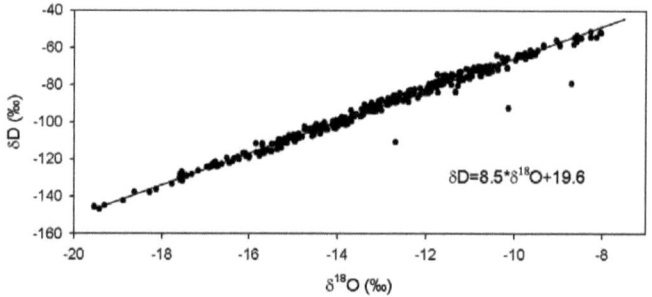

Figure 4.23: δD vs. δ^{18}O for the Colle Gnifetti, 1976–2003. The three "outliers" were not included in the calculation of the regression line.

ice core from Fiescherhorn, where d=11.0±2.0‰ (T. Jenk, personal communication). There, no correlation between d and $\delta^{18}O$ is obvious for most of the years (r^2=0.0214, 1991–2001). The behavior of d-excess can vary strongly with different climatic boundary conditions. An ice core from the Tien Shan mountains, Kyrgyzstan, Central Asia, showed an average of d=23.0‰ and a very pronounced seasonal amplitude of 15–20‰, with highest values in summer (Kreutz et al., 2003). There, the huge variability was ascribed to seasonal changes in the moisture source, transport and recycling. Using this argument for Colle Gnifetti and Fiescherhorn would mean that there is much less influence of seasonally varying moisture sources there. However, Colle Gnifetti accumulates during summer season only.

On Tsast Ula ice cap, Western Mongolia, average d-excess values of ∼15‰ and an inverted seasonality (high d during winter) were observed. These high values in a very continental site were suggested to be due to precipitation originating from re-evaporated moisture, enriched in deuterium (Schotterer et al., 1997). The same explanation might also be valid for Huascarán, Peru, where d=15.5‰ was the observed Holocene value (Thompson et al., 1995). There, the moisture source is the Atlantic Ocean and re-evaporation takes place over the Amazon basin.

In Antarctica, deuterium excess is generally much lower. A value of d=8.3‰ was reported for the Holocene period in an ice core from Dome C and d=4‰ for the Late Glacial Stage (Jouzel et al., 1982). There, this difference was interpreted in terms of higher relative humidity near the ocean surface by the end of the last glacial. In Greenland, on the other hand, the value d was found to be ∼9‰ during the last Millennium inferred from the GRIP core and core S93 (Hoffmann et al., 2001).

The deuterium excess recorded at the Swiss meteorological stations is generally lower than on Colle Gnifetti and on Fiescherhorn. The average values (based on monthly data for 1983–2004, but the data sets for deuterium are all incomplete) are shown in Table 4.7. The deuterium excess does not show significant correlation with $\delta^{18}O$ at any of the stations.

4.3.6 Calibration of stable isotopes for the use as paleo thermometer

When stable isotopes are used for paleo-thermometry for a specific site they have to be calibrated versus temperature. For that purpose, three strategies exist: Monthly resolved isotope and temperature data exist for five stations in Switzerland. There, the $\delta^{18}O$-temperature relationship was quantified and then used for interpreting the ice core record. The second strategy is to

Table 4.7: Deuterium excess at Swiss stations.

Station	Altitude	d-excess (‰)	# data
Bern	511 m	8.5±3.7	273
Grimsel	1950 m	11.3±3.3	143
Meiringen	632 m	7.7±3.3	263
Locarno	379 m	7.1±5.1	178
Gutannen	1055 m	8.4±3.8	143

compare yearly averages for Colle Gnifetti isotopes with on-site temperatures. No meteorological station, that could provide such a record, is on the glacier but temperature values that were estimated for specifically that region could be taken. If good correlation between $\delta^{18}O$ and temperature was found it would allow for a straight-forward calibration. A third strategy could be used if the correlation was poor but the same trend was observed in the isotope and temperature data. Then the different slopes of the regression curves would give another estimation for the $\delta^{18}O$-temperature relationship.

Calibration based on station data

Isotope and temperature data from five Swiss stations (Bern, Grimsel, Gutannen, Locarno and Meiringen) were taken. The stations of Bern, Grimsel, Gutannen and Meiringen are on the northern side of the main Alpine crest whereas Locarno is the only site on the southern rim of the Alps. Data for 1970–1992 were obtained from IAEA[4] and the data for 1993–2004 was provided by the Swiss Federal Office for Water and Geology[5]. For each station, summer (June, July, August) monthly $\delta^{18}O$ values were plotted against monthly temperature and linear regression was applied yielding a function $\delta^{18}O = kT - d$ which is shown in Figure 4.24. For the five stations the slope k varied between 0.47 and 0.89 ‰/°C. The average value was 0.67±0.19‰/°C. The inverse value (1.6±0.5°C/‰) is an estimate for the temperature anomaly. The coefficient of 0.67 is very close to the value of 0.69 that was originally suggested by Dansgaard (1964) for the $\delta^{18}O$-temperature relationship. This value has already been used in numerous studies to reconstruct past temper-

---

[4]IAEA/WMO (2001). Global Network of Isotopes in Precipitation. The GNIP Database. Accessible at http://isohis.iaea.org (February, 2006)

[5]Bundesamt für Wasser und Geologie in cooperation with the Institute for Physics at University of Berne, Switzerland, and the Institute for Mineralogy and Geochemistry at University of Lausanne, Switzerland.

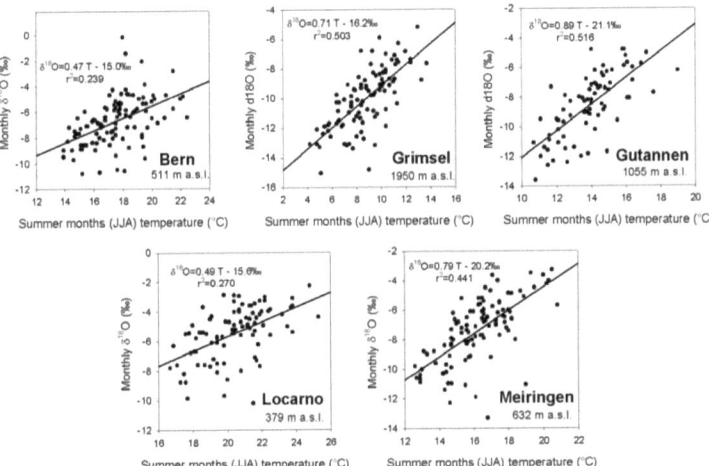

Figure 4.24: Linear relationship between $\delta^{18}O$ and air temperature for individual summer months June, July and August at five stations in Switzerland for the time period 1970–2004 A.D.

atures (Rozanski et al., 1993).

Calibration based on estimated on-site temperatures

In order to get an estimation of the temperatures on the glacier, surface air temperature anomalies from the Climate Research Unit (CRU, Jones and Moberg (2003))[6] were taken. This data set provide monthly temperature anomalies for 5° by 5° grid-points over the entire globe for the time period 1851 to present. This means that the resolution of this data set is very coarse. Here, only the temperatures for June, July and August (JJA) were taken (gridpoint 45–50°N and 5-10°E).

Another estimation of the on-site temperature was based on temperature reconstructions by Casty et al. (2005)[7] who provided a temperature reconstruction for the Alps based on a 0.5° by 0.5° grid (43.25–48.25°N and 4.25–16.25°E) and monthly resolution. For the time period 1900–2000 A.D. Casty et al. took the data from Mitchell and Jones (2005). For the comparison here, the temperature series from the grid point 45.75°N and 7.75°E was taken.

In order to perform a calibration based on annual averages, highly resolved data is needed. Therefore the calibration was limited to the time period 1976–2000 A.D., where highly resolved isotope data exists for Colle Gnifetti and the upper limit derived from the time period of the Casty et al. data ending in 2000 A.D. Figure 4.25 shows the resulting scatter plots. The CRU data is given in temperature anomaly. The Casty et al. data does not represent temperature at the high elevation site of the glacier, but as the question was the relation (slope) between $\delta^{18}O$ and temperature there is no need for adjusting the temperatures to the high elevation.

Generally, the correlation between temperature and $\delta^{18}O$ is poor ($r^2 < 0.1$). This is probably because the preservation of snow that falls during different seasons has a big influence on $\delta^{18}O$ of a single year. Other reasons for the poor correlation might be due to dating errors which would destroy the correlation or the gridded temperatures might not be a good representation of local temperatures.

Figure 4.26 shows the observed trend for Colle Gnifetti $\delta^{18}O$, Casty et al. temperatures and CRU temperature anomalies. Good correlation (r^2=0.83) is observed between the two temperature series. Between Colle Gnifetti and the temperature data, agreement is only seen for some years (1980, 1991)

---

[6]http://www.cru.uea.ac.uk/cru/data/temperature/ (February, 2006, free download). The file used was crutem2.dat

[7]http://www.ncdc.noaa.gov/paleo/pubs/casty2005/casty2005.html (March 2006)

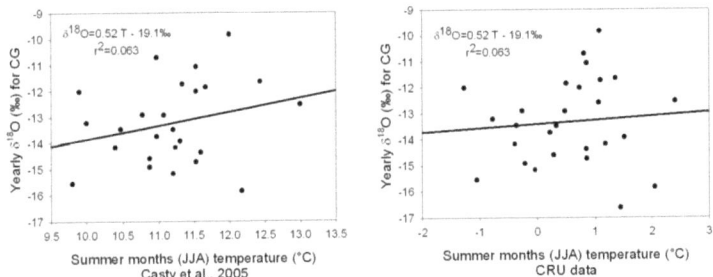

Figure 4.25: Colle Gnifetti δ^{18}O yearly averaged vs. estimated on-site summer temperature. For the CRU data anomalies are given.

such that this approach of calibrating is not very promising for the moment.

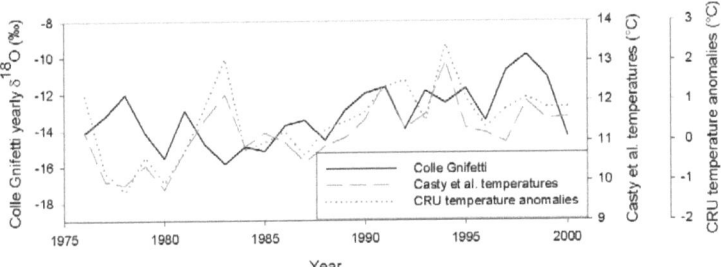

Figure 4.26: Comparison of Colle Gnifetti δ^{18}O, Casty et al. temperatures (45.75°N and 7.75°E) and CRU temperature anomalies (45–50°N and 5-10°E) 1976–2000 A.D.

Calibration by comparison of trends observed in temperature and stable isotopes

Figure 4.27 shows the observed trend (1976–2000 A.D.) in δ^{18}O and the estimated temperature (CRU and Casty et al.) on Colle Gnifetti. All three charts show a clear increase: 0.116‰ per year for δ^{18}O, 0.058°C per year for

Casty temperatures and 0.064°C per year for CRU temperature anomalies. When the slopes of the regression curves are taken, values of 2.0‰ per °C $\left(\frac{0.116}{0.058}\right)$ and 1.8‰ per °C $\left(\frac{0.116}{0.064}\right)$ are obtained for the Casty and the CRU data respectively.

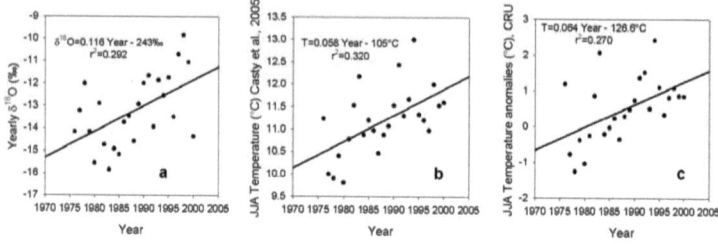

Figure 4.27: Comparison of trends in (a) δ^{18}O (annual averages for the 2003 core) and estimated temperature on Colle Gnifetti using (b) Casty et al. temperature and (c) CRU temperature anomalies.

These values appear very high especially when compared to the value of 0.69 commonly used for paleo climate reconstruction (Rozanski et al., 1993). The reason is that the increase in δ^{18}O has been much larger for 1976–2000 A.D. than the increase in estimated on-site temperatures. Maybe the temperature estimations are underestimated at such a high-elevation site they were originally not designed for. The long-temperature fluctuations on such a glacier might be larger than fluctuations at low elevation.

Keck (2001, p. 82) used a comparable approach by comparing smoothed δ^{18}O values (from KCS, KCH and Chemistry Core) and estimated on-site temperatures from the ALPCLIM gridded data set (Böhm et al., 2001). The smoothed values showed a correlation of 0.9. He came up with a sensitivity of 1.7‰/°C. However, the approach is problematic as the correlation depends on how the parameter for the smoothing is set, as stronger smoothing will produce a better correlation. An eventual underestimation of on-site temperatures is not mentioned there.

For the moment, the coefficient derived from the Swiss stations seems to be the best choice for transferring δ^{18}O into temperature, because the method works well and a strong temperature dependency of δ^{18}O in precipitation is observed north and south the main Alpine crest. However, the calibration is not straight-forward. When changes in temperature are discussed, the huge uncertainty of the coefficient (0.5–2.0‰ per °C) has to be considered.

4.3.7 Influence of shifts in seasonality of accumulation on $\delta^{18}O$

As the annual mean isotopic value strongly depends on the seasonal distribution of precipitation, it was tried to evaluate an eventual shift in the precipitation pattern on the isotope signal: In the following example, equal accumulation throughout the cold and the warm season of the year was assumed (which is not the case for Colle Gnifetti). The resulting shape of the $\delta^{18}O$ signal would then be sinusoidal, following the seasonal temperature variation. This pattern is also observed in reality and illustrated in Figure 4.28 for the station at Grimsel. Moreover, an amplitude of 5‰ (which is a typical

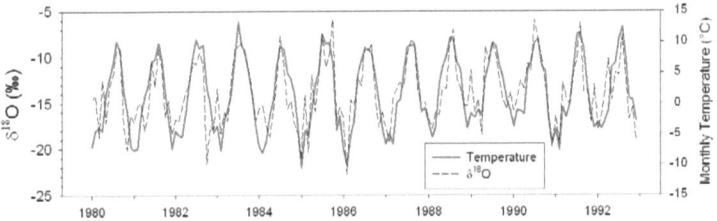

Figure 4.28: $\delta^{18}O$ and temperature variability at Grimsel station (46°34'12"N, 8°19'48"E, 1950 m a.s.l.) 1980–1993. The isotope signal has a sinusoidal shape following seasonal temperature variations.

value observed at Grimsel station) and an offset of -18‰[8] were assumed. Mathematically, the signal is then defined by $x \in [0,1]$, $y = 5\sin(2\pi x) - 18$. Homoscedastic noise[9] of ± 1 was added. The resulting signal based on 51 equidistant points ($x \in [0,1]$, $\Delta x = 0.02$) is illustrated in Figure 4.29 (solid line). The mean value of the signal is given by its integral according to the mean value theorem. It is equal to -18:

$$\bar{y} = \int_0^1 (5\sin(2\pi x) - 18)dx = -18$$

For the signal shown in Figure 4.29 (solid line), a mean of -18.00‰ was calculated.

---

[8]The offset represents the mean isotopic value, that is expected at this altitude, as discussed in Section 4.3.5

[9]Homoscedastic means that the noise is independent of the signal, whereas heteroscedastic noise is proportional to the signal

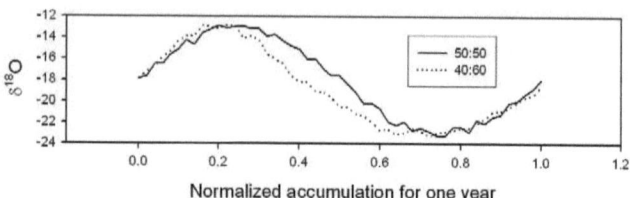

Figure 4.29: Simulated $\delta^{18}O$ variability for one year of accumulation. A mean of -18‰ and an amplitude of 5‰ were assumed. The solid line shows a signal, where 50% of the accumulation takes place during the warm season and 50% during the cold season, the dotted line assumes a 40:60 contribution of the warm season and the cold season to the accumulation, respectively.

When the pattern is changed such that 40% of the precipitation accumulate during the warm season and 60% during the cold season (which is quite a big shift), then the signal could be described by two separate functions. One sin-curve has an increased frequency such that half a cycle is completed in $x \in [0, 0.4]$. Thus, the frequency ω can be obtained: $\omega = \frac{1}{2\times 0.4} = \frac{5}{4}$. The second part of the signal is defined by a sin-curve of decreased frequency such that half a cycle is completed in $x \in [0.4, 1]$. In this case $\omega = \frac{1}{2\times 0.6} = \frac{5}{6}$. Overall, the curve is defined by $x \in [0, 0.4]$, $y = 5\sin(2\pi \frac{5}{4}x) - 18$ and $x \in [0.4, 1]$, $y = 5\sin(2\pi \frac{5}{6}x) - 18$. (Fig. 4.29, dotted line). The mean value can again be calculated by the integral:

$$\bar{y} = \int_0^{0.4} \left(5\sin\left(2\pi\frac{5}{4}x\right) - 18\right) dx + \int_{0.4}^{1} \left(5\sin\left(2\pi\frac{5}{6}x\right) - 18\right) dx =$$

$$-18 - 5\frac{2}{5\pi} = -18.64$$

The resulting mean of $\delta^{18}O$ would accordingly decrease by 12.7% ($\frac{2}{5\pi}$) of the amplitude, that is 0.64‰. The same concept is true for the opposite direction: If the pattern is changed such that 60% of the precipitation accumulate during the warm season and 40% during the cold season the mean $\delta^{18}O$ value would accordingly increase by 0.64‰ to -17.36‰.

The mean of all the values of the signal shown in Figure 4.29 (dotted line) was -18.71‰ and thus in agreement with the mathematical prediction. The difference of 0.07‰ is random and attributable to the noise that had been added to the signal.

In a more general way, any shift in the in the accumulation distribution can be expressed by φ being the fraction of precipitation during the warm season. If A is the amplitude then the curve is described by $x \in [0, \varphi]$, $y = A\sin(2\pi \frac{1}{2\varphi}x)$ and $x \in [\varphi, 1]$, $y = A\sin(2\pi \frac{1}{2(1-\varphi)}x)$. Integration gives:

$$\bar{y} = \int_0^\varphi \left(A\sin\left(\frac{2\pi}{2\varphi}x\right)\right) dx + \int_\varphi^1 \left(A\sin\left(\frac{2\pi}{2(1-\varphi)}x\right)\right) dx =$$

$$\frac{2A\varphi}{\pi} - \frac{2A(1-\varphi)}{\pi} = \frac{4A\varphi - 2A}{\pi}$$

$\bar{y}$ expresses the change in $\delta^{18}O$ as a result of a shift between precipitation during the warm and precipitation during the cold season. This relation can also be used for extreme cases, e.g. when there is no warm season accumulation any more ($\varphi = 0$). Accordingly, $\delta^{18}O$ would shift by $A\frac{-2}{\pi}$ or -3.18‰ for this example.

Estimating the on-site accumulation pattern and influences of potential shifts in the precipitation pattern

Based on the mean value of -13.1‰ it was tried to estimate the current accumulation pattern on site. Therefore, several assumptions were required. First, a yearly average of -17.6‰ for the precipitation was assumed based on the value of -17.1‰ for Grenzgletscher at 4,200 m a.s.l. (1970–1994, Eichler (2000)) using a laps rate of 0.2‰/100m, as discussed earlier. Then, an amplitude of 6‰ was used, as observed for average years on Grenzgletscher, where it generally varied between 4.5‰ and 8.0‰. Moreover, it was assumed that accumulation takes places centered at the time of the year when the highest values for $\delta^{18}O$ occur (i.e. in summer). In order to match the observed value of -13.1‰ the accumulation would have to be limited to 40% of the year centered at where maximum values occur (see Figure 4.30a).

This method of estimating the precipitation pattern on site is relatively sensitive to the assumptions it is based on, especially when the accumulation is limited to a brief fraction of the year. If the seasonal amplitude of $\delta^{18}O$ is reduced from 6‰ to 5‰ then the accumulation would have to be restricted to 26% of the year (instead of 40%) in order to explain an average value of -13.1‰. On the other hand, accumulation could take place for 60% of the year in order to yield an average value of -15.1‰.

4.3.8 Stable isotopes as temperature proxy

When isotopes are interpreted in terms of temperature, this can only be done in terms of summer temperature because the signal at Colle Gnifetti

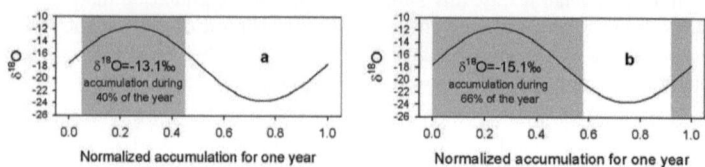

Figure 4.30: Simulated precipitation pattern using an annual mean of -17.6‰ and an amplitude of 6‰. If accumulation takes only place during the time indicated by the grey bar, the mean value of $\delta^{18}O$ is -13.1‰ (a) and -15.1‰ (b), respectively.

does not represent a yearly average. The mean $\delta^{18}O$ value of a year also strongly depends on the fraction of snow preserved during colder seasons, as a higher fraction of winter snow preservation shifts the mean towards more depleted (more negative) values. Thus, it is a priori impossible to distinguish any shift towards higher winter accumulation from a trend to lower summer temperatures.

First, $\delta^{18}O$ is interpreted as a signal for summer temperature variations, then an influence of an eventual shift in the precipitation/accumulation pattern is discussed.

High frequency oscillations are assumed to represent seasonal temperature variations although the seasons are only partially recorded, whereas the multi-decadal trend is assumed to represent long term changes in the summer temperature.

Figure 4.31 shows a linear regression over the last 103 years (1900–2003). The temporal resolution of the data is inhomogeneous because only the period 1976–2003 is available at high sampling resolution (12 samples per year). Any trend calculation with varying temporal resolution would be biased, as areas of high data density have a stronger influence on the resulting curve. Therefore, equidistant, interpolated values (14 points per year) were calculated over the time frame of interest (∼540–2003 A.D.).

These interpolated values were used for linear regression for the time period 1900–2003. The resulting line is $\delta^{18}O = 0.0206 \pm 0.002 \times Year - 54.2$, $p < 0.0001$. This means that for 1900–2003 the increase in $\delta^{18}O$ was 2.1±0.2 ‰ (from -15.1‰ to -12.9‰). This corresponds to a temperature increase of 3.4±1.1°C when the mean temperature sensitivity of 0.67‰ per °C, observed at the Swiss meteorological stations, is used. However, when the full range of the obtained temperature sensitivities is used (0.5–2.0‰ per °C) the uncertainty is much larger and the corresponding temperature increase is between

1.1 and 4.2°C.

If the entire increase in δ^{18}O from -15.1‰ (1900 A.D.) to -12.9‰ (2003 A.D.) was to be attributed to a shift in the precipitation pattern then in 1900 A.D. accumulation must have taken place during 66% of a year, as is shown in Figure 4.30b, p. 112. This would correspond to an enormous shift in the accumulation or precipitation pattern and it is regarded unlikely that such a shift accounts entirely for the observed increase. However, a fraction of the increase could indeed be due to a change in the pattern.

The observed increase was compared with surface air temperature anomalies from the Climate Research Unit. Over the land area to the south of Colle Gnifetti (5–10°E, 40–45°N), a temperature increase of 2.3°C was observed for 1900-2003 for June, July and August (JJA) according to CRU data. Within the error (3.4±1.1°C) the two values from Colle Gnifetti and CRU are similar.

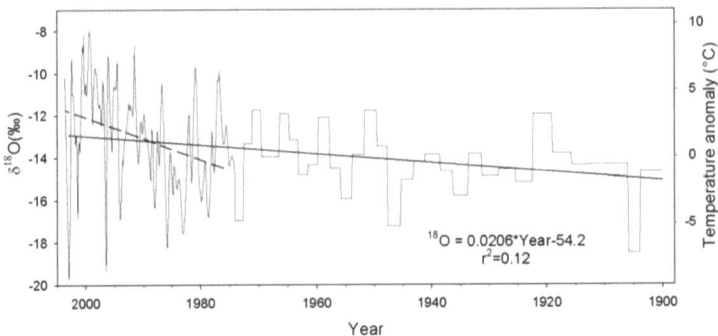

Figure 4.31: δ^{18}O record 1900–2003 A.D. The solid line shows the regression over the time period 1900–2003 based on interpolated δ^{18}O (14 points per year). The dashed line shows the regression for 1976–2003 suggesting that the warming might have intensified during that time.

4.3.9 Trends over the last 1,500 years

In the isotope data (550–2003 A.D.), the seasonal high-frequency variability is the dominant feature superimposed on a low-frequency multi-decadal trend of much lower amplitude. In order to make this trend obvious, smoothing algorithms were applied. Most of them offer an option to vary the strength of the smoothing. In extreme cases, the original data is perfectly reproduced (no smoothing) or the smoothed curve is reduced to the mean value (or linear trend) of the data set (full smoothing). Both cases are normally not useful and in practice smoothing does something in between. The "stronger" the smoothing is the more the amplitudes are reduced. This means that the smoothed data can generally not be used for quantitative interpretation or calibration.

The influence of different smoothing algorithms on the signal

All smoothing was done with the interpolated data to ensure homogeneity in the temporal data density. Three smoothing algorithms were compared: Smoothed spline functions, adjacent (moving) average and a FFT (Fast Fourier Transformation) filter. For the smoothed spline function, the smoothing factor was set to 0.80 where 1 would correspond to perfect reproduction of the data and 0 to a simple linear regression over the entire data set. The simplest smoothing, adjacent (moving) average, was tried next. In this case an average was calculated over 500 data points (∼32 years). Then, an FFT filter (350 points) was also applied[10]. The results of the three algorithms are shown in Figure 4.32. Generally, all these three algorithms lead to

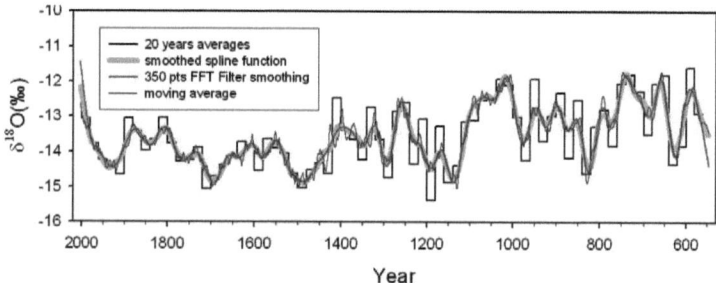

Figure 4.32: Comparison of the effect of different smoothing algorithms on the resulting signal. Adjacent average, smoothed spline functions and FFT filtering were applied. The algorithms all yield similar results such that the individual curves are hard to distinguish.

a similar result. Twenty years averages are also shown for comparison. The

---

[10]The adjacent average and the FFT filter calculations were performed with *Origin 7.5*. The smoothed spline function was calculated with a MathPack a package for Borland Dephi 7.0, http://www.lohninger.com/mathpack.html (February, 2006)

curve obtained by adjacent average is quite noisy, which does not change significantly, when the number of points, over which a value is calculated, is doubled. The FFT filter and the smoothed spline function yield almost identical curves despite the very different mathematical procedures, they are based on. The only obvious differences arise in the extremes (ends) where the FFT curve shows slightly higher amplitudes. However, the rims of smoothed data should not be over-interpreted, anyway. The generally increasing amplitude at older ages might be an artefact due to the decreasing sampling resolution there decreasing the physically available values for averaging. Finally, the smoothed spline function was used for the following discussion.

Application and discussion of the smoothed curve

When the entire high-resolution record (550–2003 A.D.) is plotted against time (Fig. 4.33), the difference between the multi-decadal and the high-frequency variability become obvious. The raw data (high-frequency) is plotted in grey whereas the thick black line is a smoothed cubic spline function indicating the "trend" in the data.

Gradual warming is observed through the 20th century (after 1935 A.D.), strongly intensifying between 1980 and 2003, which is in agreement with meteorological data (see Figure 4.34). Between 1450 (+57, -66)[11] and 1800 (+17, -20), isotope levels were generally below average. This time period is generally called the *Little Ice Age*, e.g. defined by Lamb (1969) to have occurred from 1450–1850 A.D. Around 1020 A.D., the highest $\delta^{18}O$ values of this record (-7.6‰ & -7.7‰) were observed in two samples covering about a 5-year time period and surrounded by other unusually high values. An average $\delta^{18}O$ was calculated over a 30-year period (1005–1034 A.D.) centered at this anomaly and yielded -11.7±2.3. A t-test was applied showing that this was significantly higher than the average for the recent 27-years (1976–2003) period: -13.1±2.4. Very high values were also observed around 740 A.D. This suggests that the Medieval Warm Period might have been even warmer than the last two decades of the 20th century.

4.3.10 Comparison with temperature observations and other millennial records

In a first step, the obtained $\delta^{18}O$ record was compared with air temperature anomalies for June, July and August, obtained from the data set of the

---

[11]The numbers in brackets represent the dating error: 1450 (+57, -66) means 1450 A.D. with an uncertainty ranging from 1384 A.D. to 1507 A.D.

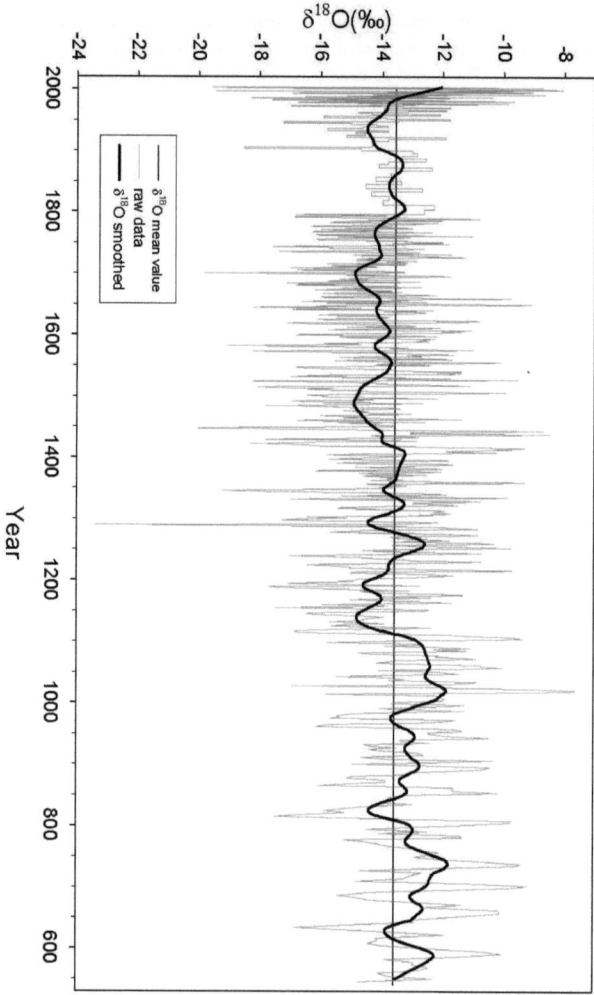

Figure 4.33: $\delta^{18}O$ record 550–2003 A.D. The thin line represents raw data with decreasing temporal resolution, the thick line represents a smoothed spline function. The straight line represents the mean of the interpolated data (-13.6‰). The section between 1800 and 1976 A.D. was sampled at coarse resolution.

Climate Research Unit for the grid points 5–10°E, 40–45°N and 5–10°E, 45–50°N, respectively. Because both data sets are very "noisy", they were smoothed using cubic spline functions. Because the smoothing parameters strongly influence the amplitude of the signal, the absolute numbers of the smoothed curves should not be directly compared. Smoothed curves serve as a guide to make patterns clearer. Figure 4.34 shows the three data sets.

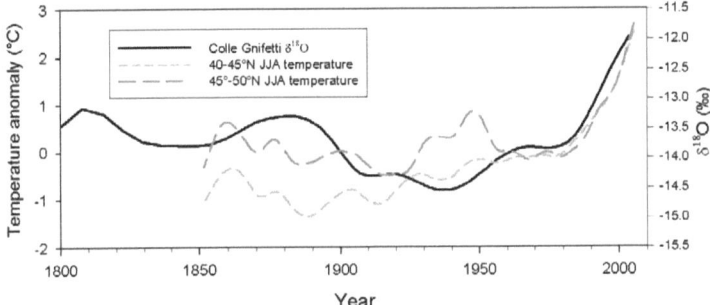

Figure 4.34: Monthly surface temperature anomaly 5–10°E, 40–50°N based on data compiled by the Climate Research Unit and Colle Gnifetti $\delta^{18}O$ record.

The agreement for 1970–2003 is quite remarkable especially the steep increase after 1980. For 1900–1970 there is some agreement between the Colle Gnifetti $\delta^{18}O$ values and the temperature of the southern grid point, before 1900, there is disagreement.

The isotope curve was also compared to temperature data provided by Casty et al. (2005), explained in Section 4.3.6. The grid-point closest to Colle Gnifetti was chosen (45.75°N and 7.75°E). Figure 4.35 shows the comparison between summer temperature at the specific site and Colle Gnifetti $\delta^{18}O$. There is little agreement in the data except that the both data sets show a strong increase after 1980.

Several millennial temperature records have been published in the recent past. Maybe the first important and most widely cited (including an important contribution to the IPCC report, IPCC - The Scientific Basis (2001, p. 134)) reconstruction was the famous "hockey stick" by Mann et al. (1999) suggesting that strong warming observed during the 20th century had been absolutely unique during the last 1,000 years. Today, more reconstructions are available and the pattern of temperature variability is apparently more complex.

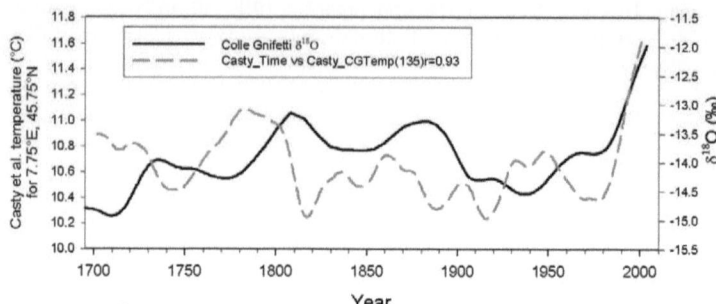

Figure 4.35: Comparison of Colle Gnifetti δ^{18}O and Casty et al. temperatures for 45.75°N and 7.75°E.

Figure 4.36 shows an overview over different reconstructions including the Mann et al. reconstruction. Obviously, the other authors find stronger variability (higher amplitude) than Mann et al. and, at first glimpse, there seems to be little agreement. However, in the high-frequency variability, there is remarkable agreement, i.e. local minima and maxima coincide well. Moreover, looking at the ensemble of all the curves, temperatures seem to have been higher for 1000–1300 A.D. (Medieval Warm Period) and after ∼1940, and temperatures were lower for 1450–1700 A.D. or, depending on whether the local maximum at ∼1780 A.D. is included or not, for 1450–1850 A.D. This picture is relatively consistent except maybe for the curve by Esper et al. (2002)) that indicates relatively low values for 1150–1350 A.D. What all the reconstructions agree about, when they are compared to instrumental data, that goes back to ∼1850 A.D., is that the 1990s were the warmest decade of the last Millennium (Kerr, 2005). The smoothed version of the Colle Gnifetti isotope record is shown for comparison. As mentioned earlier, the highest values for δ^{18}O are found around 1020 A.D. during the Medieval Warm Period and not in the last decades of the 20th century. Otherwise, there is good agreement in the main features between the reconstructions and Colle Gnifetti δ^{18}O. Note, that the dating error related to the Colle Gnifetti record is relatively big when compared to tree ring studies that are accurately dated.

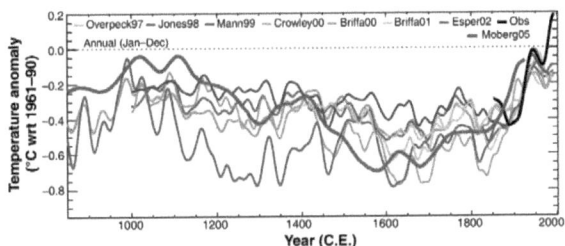

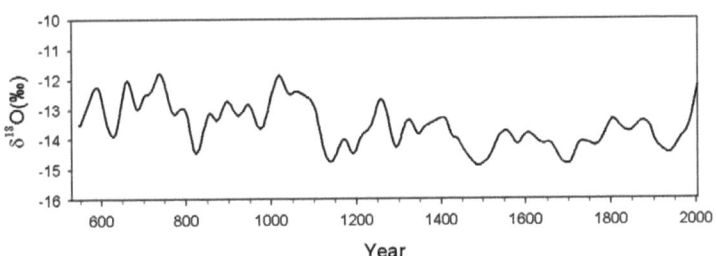

Figure 4.36: Comparison of different temperature reconstructions of the last Millennium. Temperature records recovered from tree rings and other proxies broadly agree that no time in the past millennium has been as warm as recent decades (black). Adapted from Kerr (2005). The lower figure shows smoothed $\delta^{18}O$ data from Colle Gnifetti.

4.3.11 Major ion records

Figure 4.37 shows the record of major ions in the Colle Gnifetti 2003 core. The section between 10.9 and 38.4 m weq. has not yet been analyzed. The section between 0 and 10.9 m weq. corresponds to the time period 1976–2003. According to the dating (2-parameter model), 38.4 m weq. correspond to the year 1796 (+17, -20) A.D. Thus, the lower section (38.4–62.5 m weq.) is entirely attributable to the preindustrial time. The difference between the two periods that have been analyzed is quite striking. Concentrations for some ions are by far higher during 1976–2003 than in the preindustrial time. This is true for ions that are influenced by anthropogenic emissions.

Anthropogenic sulphate e.g. mainly comes from fossil fuel burning. Coal and oil contain sulfur that is released as SO_2 and further oxidized to sulphate (SO_4^{2-}) in the troposphere. Aerosols containing sulphate are then deposited, mainly by wet deposition, on the glacier. Sulfur dioxide emissions were steadily increasing until the 1970s when filters started to be implemented in coal power plants and sulfur started to be removed chemically from liquid fuels. Since then, SO_2 emissions have been declining again. For 1976–2003 the median sulphate concentration in the core is 9.55 μeq.L^{-1}, whereas for the preindustrial age the median is 1.29 μeq.L^{-1}.

The situation is a little different for nitrate, whose precursors are nitrous oxides, NO_x. These gases are a byproduct of high temperature combustion, when they are formed in an endothermic reaction between oxygen and nitrogen. Emissions were increasing rapidly until catalytic converters for cars were introduced in the 1980s. By transferring nitrous oxides back to nitrogen and oxygen, these converters remove a big fraction of NO_x from the exhaust gases. However, this effect has been counterbalanced to some extent by steadily increasing amounts of traffic such that no reduction was obvious for the last years in this record. For 1976–2003 the median nitrate concentration in the core is 4.16 μeq.L^{-1}, whereas for the preindustrial age the median is 1.12 μeq.L^{-1}.

The only cation that is markedly influenced by an "anthropogenic increase" is ammonium (NH_4^+), where agricultural activity like the use of fertilizers containing high amounts of NH_4^+ is the main source. The release is highest during the growing season. For 1976–2003, the median ammonium concentration in the core is 6.32 μeq.L^{-1}, whereas for the preindustrial age the median is 1.36 μeq.L^{-1}.

Other ions like chloride or sodium do not show an obvious difference between the industrial and the pre-industrial age. The main source for these species has remained with sea salt.

Table 4.8 summarizes the median concentrations and compares them to

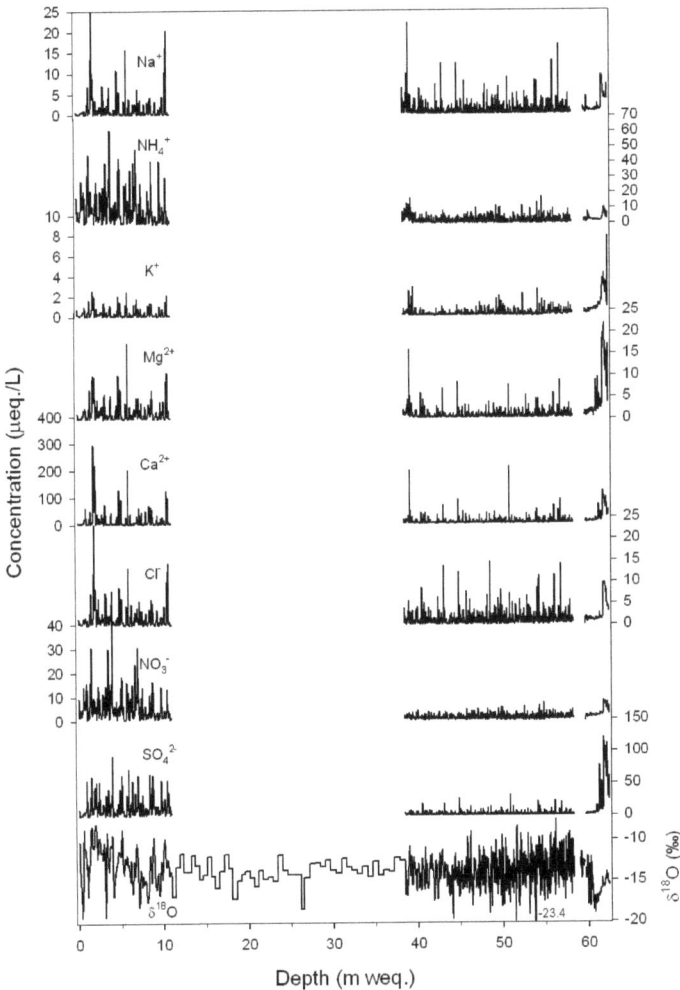

Figure 4.37: Major ions and $\delta^{18}O$ record over the entire core length. The section between 10.9 and 38.4 m weq. was only analyzed for stable isotopes.

the data from the Blue Core (Döscher, 1996). For preindustrial times there is generally good agreement between the two cores. The industrial period is defined by 1965–1981 A.D. for the Blue Core (drilled in 1982) and by 1976–2003 A.D. for the 2003 core (where chemical data is available). For nitrate and ammonium higher values are observed for the 2003 core, possibly because of increasing emissions. For sulfate a decrease would be expected as emissions have decreased over the last two decades. The median sulfate concentration is higher, though. However, sulfate is also influenced by the input of Saharan dust. Excess-sulfate can be calculated by subtracting a fraction attributable to dust and to sea salt (Schwikowski et al., 1999). The median concentration of excess-sulfate, industrial period, is 6.57 μeq./L for the 2003 core slightly lower than 7.27 μeq./L for the Blue Core.

Table 4.8: Comparison of median ion concentration (μeq./L) between the 2003 core and the Blue Core. Industrial refers to 1976–2003 for the 2003 core and to 1965–1981 for the Blue Core. Preindustrial refers to older than ~1800 for both cores.

Core	Time period	NH_4^+	SO_4^{2-}	NO_3^-
CG 2003	industrial	6.32	9.55	4.16
CG 2003	preindustrial	1.36	1.29	1.12
Blue Core	industrial	2.89	8.54	2.44
Blue Core	preindustrial	1.93	1.69	1.27

4.3.12 An intense yellow dust layer near bedrock

The most puzzling feature observed in visual stratigraphy was an intense yellow-orange dust layer near bedrock. It extends between 61.75 and 62.28 m weq. (width: 53 cm weq.) and it ends 17 cm weq. above bedrock (see Fig. 4.38 & 4.39). A similar layer was not observed in core 2.

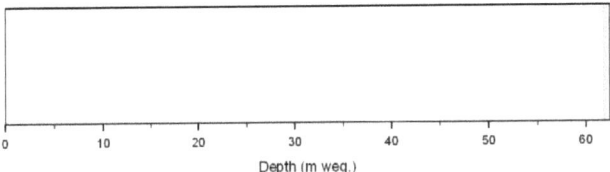

Figure 4.38: Position of the intense yellow dust layer near bedrock.

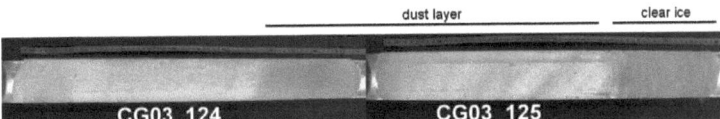

Figure 4.39: Pictures of the two lowermost core segments 124 and 125. The intense dust layer can be easily recognized. The left end of each segment is the top of the segment, therefore the clear ice at the right end of segment 125 is from the end of the ice core (bedrock).

Figure 4.40 shows the concentration profile for calcium and sulfate, the dominant ions in that section. Concentrations show a plateau at roughly 90 μeq./L throughout the visible layer. The ratio of sulfate and calcium is ~1:1, suggesting that gypsum is the main component of the water soluble fraction of the dust. In terms of mass, the concentration of $CaSO_4$ is about 6 mg/kg which is about 7% of the mass of the dust. A X-Ray Diffraction analysis (XRD) of the dust by Nicola Doebelin, Lab. chem. min. Kristallographie, University of Berne, yielded the composition: 67% illite or muscovite, 22% quartz, 7% feldspar and 3% clinochlore. Gypsum was not found in the XRD analysis as it dissolved completely when the ice was melted.

Such an intriguing yellow layer was also observed in the Red Core (Huber, 1996). Figure 4.41 shows the records of $\delta^{18}O$ and calcium in its lowermost

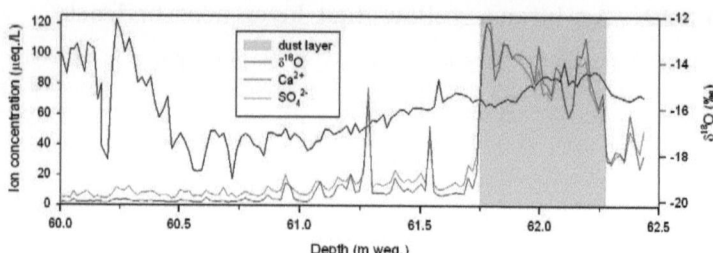

Figure 4.40: Position of the intense yellow dust layer near bedrock Calcium and sulfate concentration and $\delta^{18}O$ record around the yellow dust layer.

section. There, the layer also showed a width of ~50 cm. Calcium peaked at a concentration of 250 µeq./kg twice the maximum concentration observed in the 2003 core. After the peak, concentrations declined before the end of the core. The peak is relatively symmetric unlike the calcium (sulfate) peak in the 2003 core that shows a plateau. Sulfate and calcium have a ratio of 1:1. Huber argued that if the layer was of basal origin increasing concentration until the end of the core would be expected. In the calcium profile of both cores, concentrations show an increasing trend before the pronounced peak. $\delta^{18}O$ shows a slight decrease (-1.4‰) where the peak sits.

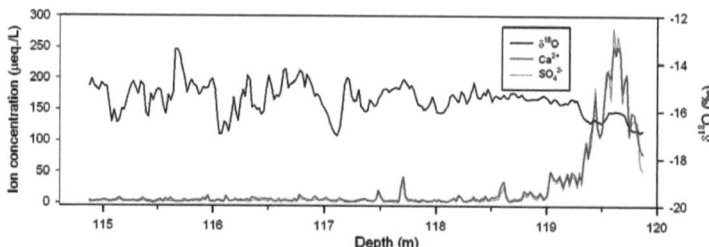

Figure 4.41: $\delta^{18}O$, calcium and sulfate concentration in the lowermost section of the Red Core. Ionic concentrations agree so well that the two curves are hard to distinguish. The data is from Huber (1996) and Schwikowski (personal communication).

This is an important difference to the 2003 core where the depletion is much higher (-4‰) and starts 1.5 m weq. before the dust layer. For both cores, the high-frequency oscillation is smoothed out where the signal starts to become depleted. The chemical fingerprint of the two layers is very similar such that they probably have the same origin. However, it is not yet understood, why ionic concentrations are about twice as high in the Red Core and why $\delta^{18}O$ looks very different.

Two possible sources are discussed: Higher dust concentration found in the Younger Dryas in Greenland around 12,000 years ago (DeAngelis et al., 1997) and some eventual influence from bedrock.

All other dust layers found in this core were attributed to input of Saharan dust. However, this particular layer needs a different explanation. Saharan dust events normally only last for a few days. At this depth in the core, annual layers have thinned to a sub-millimeter thickness therefore short events would no longer be visible or detectable. According to the dating (two-parameter model) the time span covered by this layer would be ~7,200–20,000 BP (Before Present). However, because of the extreme thinning one has to be aware of the enormous error potential in dating at these depths. The time range also formally excludes longer events like the *Younger Dryas* as the source because this very cold period at the end of the last glacial when strongly increased dust transport to Greenland resulted in higher concentrations between ~11,400–13,100 BP (DeAngelis et al., 1997) lasted for less than 2,000 years. Before dating of this section had been available, this period had been a candidate.

The fact that the layer covers the suggested time span makes its interpretation especially challenging. It would mean that the entire glacier looked as dusty as this layer for several millennia including the climatic transition from the last glacial into the Holocene. Climatic boundary conditions dramatically changed during this transition and it is difficult to imagine that such a huge shift would not have left any imprint on the composition or the intensity of the layer.

An alternative explanation was also considered. As the layer is located near bedrock, it was obvious to suggest a basal origin i.e. the layer could be the result of an interaction between the glacier and bedrock and sliding ice could have scraped of parts of the rock. Such a "dust layer" was observed at Meserve glacier, Antarctica (Cuffey et al., 2000). The glacier there showed an amber layer of dispersed fine rock particles at bedrock and a basal temperature of $-17°C$ and it could be proved that there the layer was of basal origin. The layer was in direct contact with bedrock except near the glacier tongue, where the amber layer had overridden relict ice during a glacier advance. Cuffey et al. disproved the idea that cold-based glaciers do not slide and abrade their beds. However, on Colle Gnifetti two facts argue against a basal origin:

First, the layer ends above bedrock and is followed by 19 cm of clear ice. If the dust had basal origin it must have been lifted up on top of this layer of clear ice. As the Colle Gnifetti ice core is from the glacier's uppermost accumulation zone it is unlikely that the dust layer overrode the clear ice. Second, all the mineralogical compounds found in the layer are very common for alpine rock making up the Monte Rosa massif. There, granite and granite-gneiss are the main minerals, consisting of quartz, feldspar and mica[12]. However, another compound found in the dust layer was $CaSO_4$ and this would definitely not be expected near metamorphic rock (gneiss). Geologically, calcium sulphate forms as evaporite and does not occur in rock (C. Schlüchter, personal communication). Thus, it can be concluded that the dust is of aeolian (wind-blown) origin. However, what exactly caused the layer remains so far unanswered.

---

[12]http://de.wikipedia.org/wiki/Monte_Rosa (February, 2006)

4.4 Outlook

One of the main tasks now will be the completion of the analysis of the core section 10.9–38.4 m weq. by M. Sigl, who started his PhD thesis in August, 2005. So far the interpretation of this record is hindered by this missing section where $\delta^{18}O$ is only available at a resolution of 70 cm and chemistry data is not available at all. The low resolution has an effect on the amplitude of the signal and inter-annual variability can not be determined. The time span covered is roughly 1800–1976 which is a period a lot of observational data is available. Using this section for calibration will improve the understanding of the archive. The section 1976–2003 is in many terms to brief to make a good comparison to instrumental data.

When this project was started, one of the goals was to detect some possible signals of the Roman Empire in the ice core.

The romans used a lot of lead, e.g for plumbing and they might have left some pollution imprint. According to the dating, the core should indeed cover this era. However, near 550 A.D. the core quality declines and it will be very difficult to prepare samples there avoiding contamination.

The origin of the yellow dust layer at bedrock is one of the open questions related to this core. A similar layer was also observed in the 1982 Red Core but it could not be explained there either. Is the layer related to the formation of the glacier there? Was the saddle eventually ice-free during the last glacial maximum, 21,000 years ago? Although this sounds counter-intuitive because all the great Alpine valleys were glaciated then, the cold and dry climate together with strong wind might have prevented accumulation. In Antarctica, some areas are ice-free today (e.g. Armstrong, 1978), despite extremely low temperatures there. There is just too little precipitation for the formation of glaciers.

The lower-most part of core segment 125, just above bedrock, consists of clear ice. What might the age of the ice be there? So far the sample size was considered to small (<1kg) to enable a ^{14}C determination but this may change in the future.

Chapter 5
Acknowledgements

I would like to express my deepest thanks to[1]

- Uwe & Gertraud Bolius, my parents, for their endless logistic and financial support during my entire educational career and for personal support in a complicated life.

- Sabina Brütsch for introducing me into *Ion Chromatography*, for teaching me lessons on how to taste wine for having good food and good fun, and especially for creating a good atmosphere between students and the staff at Paul Scherrer Institut.

- Anita Ciric for reinforcing the ice core group at Paul Scherrer Institute and thus contributing the a stimulating atmosphere.

- Rugard Dressler for his friendly, encouraging and clever support with all statistic-related topics.

- Jost Eikenberg for offering support and for finding the 1963 nuclear weapon horizon in the Colle Gnifetti core by analyzing samples for tritium.

- Heinz W. Gäggeler for the supervision of my work during the last three years and for always taking some time when I needed something from him despite of being very busy.

- Monika Gloor for analyzing 6 meters (250 samples) of the Colle Gnifetti ice core for major ions and $\delta^{18}O$. Although the results disagreed with what I had expected, I was happy to have them.

---

[1] In alphabetic order

- Kai Hassler for providing the excellent photo of Colle Gnifetti (page 77) but also for alpinistic adventures including rock & ice climbing and ski touring.
- Keith Henderson for sharing his impressive knowledge of paleo climatology with us, for having fun in the office and during the famous *Wine & Cheese Parties*.
- Thomas Huthwelker for encouraging support in LaTeX and for his funny jokes.
- Theo Jenk for sharing the exciting experience of field campaigns in the South-American wilderness and of course for measuring ^{14}C on samples from the Colle Gnifetti.
- Thomas Kellerhals for interesting discussions on climate dynamics, for reviewing the manuscript and for very motivating attitude in free climbing.
- Claudia Kerner for allowing me to accept a position in Switzerland and for sharing the consequence that we had to live separated during much of my stay here. And of course for assistance on this manuscript.
- Andreas Laube for his countless isotope ratio measurements and for his great effort ensuring high quality data.
- Anne Palmer for introducing me into ice cutting, for practical advice including lessons about how to use Excel and also for personal assistance during hard times in the course of my dissertation as well as for corrections of the manuscript
- Norbert Preining for LaTeX support, mathematical assistance and corrections of the manuscript.
- Matthias Saurer for helping me so much with the mass spectrometer.
- Aurel Schwerzmann for providing wonderful pictures from the Fiescherhorn 2002 drilling campaign.
- Margit Schwikowski for the supervision of my work, for offering me the opportunity of participating in exciting field work, for always allowing us to interrupt her work whenever we had a question, and, of course, for raising funds enabling my employment at Paul Scherrer Institute.

- Rolf Siegwolf for his endless instrumental support for the mass spectrometer and for fruitful and encouraging discussions about many different topics.

- Michael Sigl for reinforcing the ice core group at Paul Scherrer Institute and bringing new ideas, views and climbing power into the team.

- The Swiss National Science Foundation (Grant no. 200021-100289) for the funding of my project and for accepting the change of topic after the field campaign 2004 failure.

- Leo Tobler for the lead-210 analysis

- Edith Vogel for the chemical preparation of the lead-210 samples

- All the members of the Laboratory for Radiochemistry and Environmental Chemistry at University of Berne and at Paul Scherrer Institut.

...Thank you!

Chapter 6
Annex

6.1 Bibliography - Literature

Recommended Reading (books):

- *Earth's climate: Past and Future.* by William F. Ruddiman (2001). An excellent introduction to everything in climate research: Climate reconstruction, its dynamics and the future. This book is very illustrative and probably the most comprehensive one available at the moment. A "must" for anyone interested in the functioning of climate.

- *Klima im Wandel* by Christian-Dietrich Schönwiese (1992). Written in German. An excellent introduction to how climate works, also very interesting for non-scientists. Even easier to read than Ruddiman's book but it also goes less into details.

- *Ice ages: Solving the mystery* by J. Imbrie and K.P. Imbrie (1979). A widely read book that deals much with the history of the discovery of ice ages (their concept only emerged in the 19th century). The authors suggested that the major problem(s) had been solved, but this has since been reconsidered again (and again...). A scientific novel.

- *Earth Paleoenvironments: Records Preserved in Mid- and Low-Lattitude Glaciers.* Editors: L. DeWayne Cecil, Jaromy R. Green and Lonnie G. Thompson (2004). A good introduction to the more specific topic of ice cores recovered from non-polar glaciers.

- *The atmosphere — An Introduction to Meteorology* by Lutgens and Tarbuck (2001). An excellent book about basic meteorology. It also covers many climate-related topics, like precipitation patterns on earth,

temperature gradients, general wind and ocean circulation, El Niño . . . .
Very illustrative.

Bibliography

Aceituno, P., 1988. On the functioning of the Southern Oscillation in the South American sector. *Monthly Weather Review*, **116(3)**, 505–524.

Andersen, K. K., Azuma, N., Barnola, J. M., Bigler, M., Biscaye, P., Caillon, N., Chappellaz, J., Clausen, H. B., DahlJensen, D., Fischer, H., Fluckiger, J., Fritzsche, D., Fujii, Y., Goto-Azuma, K., Gronvold, K., Gundestrup, N. S., Hansson, M., Huber, C., Hvidberg, C. S., Johnsen, S. J., Jonsell, U., Jouzel, J., Kipfstuhl, S., Landais, A., Leuenberger, M., Lorrain, R., Masson-Delmotte, V., Miller, H., Motoyama, H., Narita, H., Popp, T., Rasmussen, S. O., Raynaud, D., Rothlisberger, R., Ruth, U., Samyn, D., Schwander, J., Shoji, H., Siggard-Andersen, M. L., Steffensen, J. P., Stocker, T., Sveinbjornsdottir, A. E., Svensson, A., Takata, M., Tison, J. L., Thorsteinsson, T., Watanabe, O., Wilhelms, F., and White, J. W. C., 2004. High-resolution record of Northern Hemisphere climate extending into the last interglacial period. *Nature*, **431(7005)**, 147–151.

Armbruster, T., 2000. *Stratigraphische Datierung hoch-alpiner Eisbohrkerne über die letzten 1000 Jahre*. Master's thesis, Institut für Umweltphysik, University of Heidelberg.

Armstrong, R. L., 1978. K-Ar Dating - Late Cenozoic Mcmurdo Volcanic Group and Dry Valley Glacial History, Victoria Land, Antarctica. *New Zealand Journal of Geology and Geophysics*, **21(6)**, 685–698.

Augustin, L., Barbante, C., Barnes, P. R. F., Barnola, J. M., Bigler, M., Castellano, E., Cattani, O., Chappellaz, J., DahlJensen, D., Delmonte, B., Dreyfus, G., Durand, G., Falourd, S., Fischer, H., Fluckiger, J., Hansson, M. E., Huybrechts, P., Jugie, R., Johnsen, S. J., Jouzel, J., Kaufmann, P., Kipfstuhl, J., Lambert, F., Lipenkov, V. Y., Littot, G. V. C., Longinelli, A., Lorrain, R., Maggi, V., Masson-Delmotte, V., Miller, H., Mulvaney, R., Oerlemans, J., Oerter, H., Orombelli, G., Parrenin, F., Peel, D. A., Petit, J. R., Raynaud, D., Ritz, C., Ruth, U., Schwander, J., Siegenthaler, U., Souchez, R., Stauffer, B., Steffensen, J. P., Stenni, B., Stocker, T. F.,

Tabacco, I. E., Udisti, R., van de Wal, R. S. W., van den Broeke, M., Weiss, J., Wilhelms, F., Winther, J. G., Wolff, E. W., and Zucchelli, M., 2004. Eight glacial cycles from an Antarctic ice core. *Nature*, **429(6992)**, 623–628.

Baertschi, P., 1976. Absolute O-18 Content of Standard Mean Ocean Water. *Earth and Planetary Science Letters*, **31(3)**, 341–344.

Baltensperger, U., Gäggeler, H., Jost, D., Lugauer, M., Schwikowski, M., Weingartner, E., and Seibert, P., 1997. Aerosol climatology at the high-alpine site Jungfraujoch, Switzerland. *Journal of Geophysical Research - Atmospheres*, **102(D16)**, 19 707–19 715.

Böhlert, R., 2005. *Glaziologische Untersuchung auf dem Colle Gnifetti und auf dem Mt. Blanc: Ermittlung der Eisdickenverteilung und interner Schichten mittels Georadar*. Master's thesis, Geographisches Institut, Universität Zürich.

Böhm, R., Auer, I., Brunetti, M., Maugeri, M., Nanni, T., and Schöner, W., 2001. Regional temperature variability in the European Alps 1760–1998 from homogenized instrumental time series. *International Journal of Climatology*, **21**, 1779–1801.

Bolzan, J. F., 1985. Ice Flow at the Dome-C Ice Divide Based on a Deep Temperature Profile. *Journal of Geophysical Research-Atmospheres*, **90(D5)**, 8111–8124.

Bradley, R., Vuille, M., Hardy, D., and Thompson, L., 2003. Low latitude ice cores record Pacific sea surface temperatures. *Geophysical Research Letters*, **30(4)**.

Casty, C., Wanner, H., Luterbacher, J., Esper, J., and Böhm, R., 2005. Temperature and precipitation variability in the european Alps since 1500. *International Journal of Climatology*, **25(14)**, 1855–1880.

Cecil, L. D., Green, J. R., and Thompson, L. G., editors, 2004. *Earth Paleoenvironments: Records Preserved in Mid- and Low-Lattitude Glaciers*, volume 9 of *Developments in Paleoenvironmental Research*. Kluwer Academic Publishers, Dordrecht/Boston/London.

Chinsamy, A., Rich, T., and Vickers-Rich, P., 1998. Polar dinosaur bone histology. *Journal of Vertebrate Paleontology*, **18(2)**, 385–390.

Clark, I. D. and Fritz, P., 1997. *Environmental Isotopes in Hydrogeology*. Lewis Publishers, Boca Raton / New York.

CLIMAP, 1981. Seasonal reconstructions of the EarthŠs surface at the last glacial maximum. Map series, technical report mc-36, Geological Society of America.

Crozaz, G. and Langway, C., 1966. Dating Greenland firn-ice cores with ^{210}Pb. *Earth and Planetary Science Letters*, **1(4)**, 194–196.

Cuffey, K. M., Conway, H., Gades, A. M., Hallet, B., Lorrain, R., Severinghaus, J. P., Steig, E. J., Vaughn, B., and White, J. W. C., 2000. Entrainment at cold glacier beds. *Geology*, **28(4)**, 351–354.

Cullen, H. M., deMenocal, P. B., Hemming, S., Hemming, G., Brown, F. H., Guilderson, T., and Sirocko, F., 2000. Climate change and the collapse of the Akkadian empire: Evidence from the deep sea. *Geology*, **28(4)**, 379–382.

Dällenbach, A., 2000. *Methan- und Lachgasmessungen aus Arktis, Antarktis und den Alpen*. Ph.D. thesis, Universität Bern.

Dansgaard, W., 1964. Stable isotopes in precipitation. *Tellus*, **XVI(4)**, 437–467.

Dansgaard, W. and Johnsen, S. J., 1969. A flow model and a time scale for the ice core from Camp Century, Greenland. *Journal of Glaciology*, **8(53)**, 215–223.

Dansgaard, W., Johnsen, S. J., Moller, J., and Langway, C. C., 1969. One Thousand Centuries of Climatic Record from Camp Century on Greenland Ice Sheet. *Science*, **166(3903)**, 377–381.

DeAngelis, M., Steffensen, J. P., Legrand, M., Clausen, H., and Hammer, C., 1997. Primary aerosol (sea salt and soil dust) deposited in Greenland ice during the last climatic cycle: Comparison with east Antarctic records. *Journal of Geophysical Research-Oceans*, **102(C12)**, 26 681–26 698.

Dhainaut, J. F., Claessens, Y. E., Ginsburg, C., and Riou, B., 2004. Unprecedented heat-related deaths during the 2003 heat wave in Paris: consequences on emergency departments. *Critical Care*, **8(1)**, 1–2.

Döscher, A., 1996. *Historische Entwicklung von atmosphärischen Spurenstoffen rekonstruiert aus Firn und Eis alpiner Gletscher*. Ph.D. thesis, Universität Bern.

Döscher, A., Gäggeler, H., Schotterer, U., and Schwikowski, M., 1996. A historical record of ammonium concentrations from a glacier in the Alps. *Geophysical Research Letters*, **23(20)**, 2741–2744.

Döscher, A., Gäggeler, H. W., Schotterer, U., and Schwikowski, M., 1995. A 130 years deposition record of sulfate, nitrate and chloride from a high-alpine glacier. *Water Air and Soil Pollution*, **85(2)**, 603–609.

Dubois, J.-D. and Flück, J., 1984. Geochemistry: utilisation of geothermal resources of the Baden area. Technical report, Swiss National Energy Research Foundation, NEFF 165-1B-032.

Eichler, A., 2000. *Deposition von Spurenstoffen in Firn und Eis alpiner Gletscher, Untersuchung von Nord-Süd-Gradienten*. Ph.D. thesis, Universität Bern.

Eichler, A., Schwikowski, M., Gäggeler, H., Furrer, V., Synal, H., Beer, J., Saurer, M., and Funk, M., 2000. Glaciochemical dating of an ice core from upper Grenzgletscher (4200 ma.s.l.). *Journal of Glaciology*, **46(154)**, 507–515.

Escobar, F. and Aceituno, P., 1998. Influencia del fenómeno ENSO sobre la precipitación nival en el sector andino de Chile Central, durante el invierno austral. *Bulletin de l'Institut Français d'Études Andines*, **27**, 753–759.

Escobar, F., Casassa, G., and Veronica, P., 1995. Variaciones de un glaciar de montaña en los Andes de Chile central en las ultimas dos decadas. *Bulletin de l'Institut Français d'Études Andines*, **24(3)**, 683–695.

Esper, J., Cook, E. R., and Schweingruber, F. H., 2002. Low-frequency signals in long tree-ring chronologies for reconstructing past temperature variability. *Science*, **295(5563)**, 2250–2253.

Esper, J., Frank, D., and Wilson, J., 2004. Climate reconstructions: Low-frequency ambition and high-frequency ratification. *Eos*, **85**, 133,120.

Fiorillo, A. R. and Parrish, J. T., 2004. The first record of a Cretaceous dinosaur from southwestern Alaska. *Cretaceous Research*, **25(4)**, 453–458.

Gäggeler, H., von Gunten, H., Rössler, E., Öschger, H., and Schotterer, U., 1983. ^{210}Pb Dating of cold alpine firn/ice cores from Colle Gnifetti, Switzerland. *Journal of Glaciology*, **29(101)**, 165–177.

Gäggeler, H. W., Jost, D. T., Baltensperger, U., Schwikowski, M., and Seibert, P., 1995. Radon and thoron decay product and ^{210}Pb measurements at Jungfraujoch, Switzerland. *Atmospheric Environment*, **29(5)**, 607–616.

Gehre, M. and Strauch, G., 2003. High-temperature elemental analysis and pyrolysis techniques for stable isotope analysis. *Rapid Communications in Mass Spectrometry*, **17(13)**, 1497–1503.

Ginot, P., 2001. *Glaciochemical study of ice cores from Andean glaciers*. Ph.d. thesis, Dep. Chem. and Biochem., University of Berne.

Ginot, P., Kull, C., Schotterer, U., Schwikowski, M., and Gäggeler, H. W., 2005. Glacier mass balance reconstruction by sublimation induced enrichment of chemical species on Cerro Tapado (Chilean Andes). *Climate of the Past Discussions*, **1**, 169–192, submitted.

Ginot, P., Kull, C., Schwikowski, M., Schotterer, U., and Gäggeler, H., 2001a. Effects of postdepositional processes on snow composition of a subtropical glacier (Cerro Tapado, Chilean Andes). *Journal of Geophysical Research - Atmospheres*, **106(D23)**, 32 375–32 386.

Ginot, P., Schwikowski, M., Schotterer, U., Stichler, W., Gäggeler, H., Francou, B., Gallaire, R., and Pouyaud, B., 2002. Potential for climate variability reconstruction from Andean glaciochemical records. *Annals of Glaciology*, **35**, 443–450.

Ginot, P., Stampfli, D., Stampfli, F., Schwikowski, M., and Gäggeler, H., 2001b. FELICS, a new ice core drilling system for high-altitude glaciers. *Memoirs of National Institute of Polar Research*, **56(Special Issue)**, 38–48.

Glueck, M. F. and Stockton, C. W., 2001. Reconstruction of the North Atlantic Oscillation, 1429-1983. *International Journal of Climatology*, **21(12)**, 1453–1465.

Götz, F. W. P., 1936. Staubfälle in Arosa im Spätwinter 1936. *Meteorologische Zeitschrift*, **53**, 227.

Hagemann, R., Nief, G., and Roth, E., 1970. Absolute Isotopic Scale for Deuterium Analysis of Natural Waters Absolute D/H Ratio for SMOW. *Tellus*, **22(6)**, 712–715.

Haug, G. H., Ganopolski, A., Sigman, D. M., Rosell-Mele, A., Swann, G. E. A., Tiedemann, R., Jaccard, S. L., Bollmann, J., Maslin, M. A., Leng,

M. J., and Eglinton, G., 2005. North Pacific seasonality and the glaciation of North America 2.7 million years ago. *Nature*, **433(7028)**, 821–825.

Haug, G. H., Gunther, D., Peterson, L. C., Sigman, D. M., Hughen, K. A., and Aeschlimann, B., 2003. Climate and the collapse of Maya civilization. *Science*, **299(5613)**, 1731–1735.

Henderson, K., Thompson, L., and Lin, P., 1999. Recording of El Niño in ice core $\delta^{18}O$ records from Nevado Huascarán, Peru. *Journal of Geophysical Research - Atmospheres*, **104(D24)**, 31053–31065.

Hoffman, P., Kaufman, A., Halverson, G., and Schrag, D., 1998. A Neoproterozoic snowball earth. *Science*, **281(5381)**, 1342–1346.

Hoffmann, G., Jouzel, J., and Johnsen, S., 2001. Deuterium excess record from central Greenland over the last millennium: Hints of a North Atlantic signal during the Little Ice Age. *Journal of Geophysical Research-Atmospheres*, **106(D13)**, 14265–14274.

Hoffmann, G., Ramirez, E., Taupin, J., Francou, B., Ribstein, P., Delmas, R., Durr, H., Gallaire, R., Simões, J., Schotterer, U., Stievenard, M., and Werner, M., 2003. Coherent isotope history of Andean ice cores over the last century. *Geophysical Research Letters*, **30(4)**.

Huber, T., 1996. *Messung anorganischer Spurenstoffe in Eis mittels Kapillarelektrophorese*. Master's thesis, Departement für Chemie und Biochemie, Universität Bern.

Imbrie, J. and Imbrie, K. P., 1979. *Ice ages : solving the mystery*. Short Hills, New Jersey Enslow, New Jersey.

IPCC - Synthesis report, 2001. Climate Change 2001: Synthesis Report. Technical report, Intergovernmental Panel on Climate Change.

IPCC - The Scientific Basis, 2001. Climate Change 2001: The Scientific Basis. Technical report, Intergovernmental Panel on Climate Change.

Jenk, T., Szidat, S., Schwikowski, M., Gaeggeler, H. W., Bolius, D., Wacker, L., Synal, H. A., and Saurer, M., 2005. Microgram Level Radiocarbon (C-14) Determination on Carbonaceous Particles in Ice. *Nuclear Instruments and Methods in Physics Research B*, **submitted**.

Jones, P. D. and Moberg, A., 2003. Hemispheric and large-scale surface air temperature variations: An extensive revision and an update to 2001. *Journal of Climate*, **16(2)**, 206–223.

Jouzel, J., Merlivat, L., and Lorius, C., 1982. Deuterium Excess in an East Antarctic Ice Core Suggests Higher Relative-Humidity at the Oceanic Surface During the Last Glacial Maximum. *Nature*, **299(5885)**, 688–691.

Keck, L., 2001. *Climate significance of stable isotope records from Alpine ice cores*. Ph.D. thesis, University of Heidelberg.

Kerr, R. A., 2005. Global warming - Millennium's hottest decade retains its title, for now. *Science*, **307(5711)**, 828–829.

King, D. A., 2004. Environment - Climate change science: Adapt, mitigate, or ignore? *Science*, **303(5655)**, 176–177.

Knüsel, S., 2003. *Continuous Trace-Element Analysis and Identification of Climate Signals using an Ice Core from Nevado Illimani, Bolivia*. Ph.D. thesis, University of Berne.

Knüsel, S., Ginot, P., Schotterer, U., Schwikowski, M., Gäggeler, H., Francou, B., Petit, J., Simões, J., and Taupin, J., 2003. Dating of two nearby ice cores from the Illimani, Bolivia. *Journal of Geophysical Research - Atmospheres*, **108(D6)**.

Kreutz, K. J., Wake, C. P., Aizen, V. B., Cecil, L. D., and Synal, H. A., 2003. Seasonal deuterium excess in a Tien Shan ice core: Influence of moisture transport and recycling in Central Asia. *Geophysical Research Letters*, **30(18)**. 1922.

Kull, C., Grosjean, M., and Veit, H., 2000. Late Pleistocene climate conditions in the North Chilean Andes drawn from a climate-glacier model. *Journal of Glaciology*, **46**, 622–632.

Lamb, H. H., 1969. Climatic fluctuations. In H. Flohn, editor, *World survey of climatology, 2, General climatology*, pages 173–249. Elsevier, New York.

Legrand, M., Preunkert, S., Wagenbach, D., and Fischer, H., 2002. Seasonally resolved Alpine and Greenland ice core records of anthropogenic HCl emissions over the 20th century. *Journal of Geophysical Research-Atmospheres*, **107(D12)**. 4139.

Lehmann, M. and Siegenthaler, U., 1991. Equilibrium oxygen- and hydrogen-isotope fractionation between ice and water. *Journal of Glaciology*, **37(125)**, 23–26.

Leiva, J., 1999. Recent fluctuations of the Argentinian glaciers. *Global and Planetary Change*, **22(1-4)**, 169–177.

Luterbacher, J., Dietrich, D., Xoplaki, E., Grosjean, M., and Wanner, H., 2004. European seasonal and annual temperature variability, trends, and extremes since 1500. *Science*, **303(5663)**, 1499–1503.

Lutgens, F. K. and Tarbuck, E. J., 2001. *The Atmosphere - An Introduction to Meteorology*. Prentice Hall, Upper Saddle River, New Jersey 07458, 8th edition.

Lüthi, M. and Funk, M., 2000. Dating ice cores from a high Alpine glacier with a flow model for cold firn. *Annals of Glaciology*, **31**, 69–79.

Mann, M. E., Bradley, R. S., and Hughes, M. K., 1999. Northern hemisphere temperatures during the past millennium: Inferences, uncertainties, and limitations. *Geophysical Research Letters*, **26(6)**, 759–762.

Markgraf, V., Baumgartner, T. R., Bradbury, J. P., Diaz, H. F., Dunbar, R. B., Luckman, B. H., Seltzer, G. O., Swetnam, T. W., and Villalba, R., 2000. Paleoclimate reconstruction along the Pole-Equator-Pole transect of the Americas (PEP 1). *Quaternary Science Reviews*, **19(1-5)**, 125–140.

Masson-Delmotte, V., Jouzel, J., Landais, A., Stievenard, M., Johnsen, S. J., White, J. W. C., Werner, M., Sveinbjornsdottir, A., and Fuhrer, K., 2005. GRIP deuterium excess reveals rapid and orbital-scale changes in Greenland moisture origin. *Science*, **309(5731)**, 118–121.

Milana, J. and Maturano, A., 1999. Application of Radio Echo Sounding at the arid Andes of Argentina: the Agua Negra Glacier. *Global and Planetary Change*, **22(1-4)**, 179–191.

Milankovitch, M., 1941. *Kanon der Erdbestrahlung und seine Anwendung auf das Eiszeitenproblem*. Royal Serb. Acad., Spec. Publ., 133, Belgrade.

Mitchell, T. D. and Jones, P. D., 2005. An improved method of constructing a database of monthly climate observations and associated high-resolution grids. *International Journal of Climatology*, **25(6)**, 693–712.

Moberg, A., Sonechkin, D. M., Holmgren, K., Datsenko, N. M., and Karlen, W., 2005. Highly variable Northern Hemisphere temperatures reconstructed from low- and high-resolution proxy data. *Nature*, **433(7026)**, 613–617.

Narod, B. and Clarke, G., 1994. Miniature high-power impulse transmitter for radio-echo sounding. *Journal of Glaciology*, **40(134)**, 190–194.

Nishimura, K. and Hunt, J., 2000. Saltation and incipient suspension above a flat particle bed below a turbulent boundary layer. *Journal of Fluid Mechanics*, **417**, 77–102.

Nye, J., 1963. Correction factor for accumulation measured by the thickness of the annual layers in an ice sheet. *Journal of Glaciology*, **4(36)**, 785–788.

Paterson, W., 1994. *The Physics of Glaciers*. Pergamon/Elsevier, Oxford, third edition.

Paul, F., Kaab, A., Maisch, M., Kellenberger, T., and Haeberli, W., 2004. Rapid disintegration of Alpine glaciers observed with satellite data. *Geophysical Research Letters*, **31(21)**. L21402.

Petit, J. R., Jouzel, J., Raynaud, D., Barkov, N. I., Barnola, J. M., Basile, I., Bender, M., Chappellaz, J., Davis, M., Delaygue, G., Delmotte, M., Kotlyakov, V. M., Legrand, M., Lipenkov, V. Y., Lorius, C., Pepin, L., Ritz, C., Saltzman, E., and Stievenard, M., 1999. Climate and atmospheric history of the past 420,000 years from the Vostok ice core, Antarctica. *Nature*, **399(6735)**, 429–436.

Preunkert, S., Wagenbach, D., and Legrand, M., 2003. A seasonally resolved alpine ice core record of nitrate: Comparison with anthropogenic inventories and estimation of preindustrial emissions of NO in Europe. *Journal of Geophysical Research-Atmospheres*, **108(D21)**, art. no. 4681. NOV 14.

Prodi, F. and Fea, G., 1978. Transport and deposition of Saharan dust over the Alps. *Proceedings der 15. Internationalen Tagung für Alpine Meteorologie, Grindelwald, Publikationen der Schweizerischen Meteorologischen Anstalt*, **40**, 179–182.

Ramirez, E., Hoffmann, G., Taupin, J., Francou, B., Ribstein, P., Caillon, N., Ferron, F., Landais, A., Petit, J., Pouyaud, B., Schotterer, U., Simões, J., and Stievenard, M., 2003. A new Andean deep ice core from Nevado Illimani (6350 m), Bolivia. *Earth and Planetary Science Letters*, **212(3-4)**, 337–350.

Rivera, A. and Casassa, G., 2002. Detection of ice thickness using radio-echo sounding on the Southern Patagonia Icefield. In: The Patagonian Icefields: A Unique Natural Laboratory for Environmental and Climate Change Studies. Eds. G. Casassa, F. Sepúlveda and R. Sinclair. Series of the Centro de Estudios Científicos. *Kluwer Academic/Plenum Publishers, New York*, **35**, 101–115.

Rivera, A., Casassa, G., and Acuña, C., 2001. Mediciones de espesor en glaciares de Chile centrosur. *Revista Investigaciones Geográficas*, **35**, 67–100.

Rozanski, K., Araguas, L., and Gonfiantini, R., 1993. Isotopic Patterns in Modern Global Precipitation. In P. K. Swart, K. C. Lohmann, J. McKenzie, and S. Savin, editors, *Climate Change in Continental Isotopic Records, Geophysical Monograph 78,*, pages 1–37. American Geophysical Union.

Ruddiman, W. F., 1997. *Tectonic Uplift and Climate Change*. Plenum Press, New York.

Ruddiman, W. F., 2001. *Earth's climate: Past and Future*. W. H. Freeman and Company, New York.

Ruddiman, W. F. and Thomson, J. S., 2001. The case for human causes of increased atmospheric CH4. *Quaternary Science Reviews*, **20(18)**, 1769–1777.

Saurer, M., Robertson, I., Siegwolf, R., and Leuenberger, M., 1998. Oxygen isotope analysis of cellulose: An interlaboratory comparison. *Analytical Chemistry*, **70(10)**, 2074–2080.

Schäfer, J., 1995. *Reconstruction of bio-geochemical cycles using an Alpine ice core*. Master's thesis, Institut für Umweltphysik, Universität Heidelberg.

Schär, C., Vidale, P. L., Lüthi, D., Frei, C., Häberli, C., Liniger, M. A., and Appenzeller, C., 2004. The role of increasing temperature variability in European summer heatwaves. *Nature*, **427**, 332–336.

Schneider, C. and Gies, D., 2004. Effects of El Niño-southern oscillation on southernmost South America precipitation at 53 degrees S revealed from NCEP-NCAR reanalyses and weather station data. *International Journal of Climatology*, **24(9)**, 1057–1076.

Schönwiese, C.-D., 1992. *Klima im Wandel*. Deutsche Verlags-Anstalt, Stuttgart.

Schotterer, U., 2004. The Influence of Post-Depositional Effects on Ice Core Studies: Examples from the Alps, Andes and Altai. In L. D. Cecil, J. R. Green, and L. G. Thompson, editors, *Earth Paleoenvironments: Records Preserved in Mid- and Low-Lattitude Glaciers*, volume 9, pages 39–60. Kluwer Academic Publishers, Dordrecht/Boston/London.

Schotterer, U., Froehlich, K., Gäggeler, H. W., Sandjordj, S., and Stichler, W., 1997. Isotope records from Mongolian and Alpine ice cores as climate indicators. *Climatic Change*, **36**, 519–530. Using Smart Source Parsing Pages 519-530 English Serial.

Schotterer, U., Haeberli, W., Good, W., Oeschger, H., and Röthlisberger, H., 1978. Datierung von kaltem Firn und Eis in einem Bohrkern vom Colle Gnifetti, Monte Rosa. In *Jahrbuch der Schweizerischen Naturforschenden Gesellschaft*, pages 48–57. Schweizerische Naturforschende Gesellschaft.

Schotterer, U., Oeschger, H., Wagenbach, D., and Münnich, K., 1985. Information on paleo-precipitation on a high-altitude glacier Monte Rosa, Switzerland. *Zeitschrift fuer Gletscherkunde und Glazialgeologie*, **21(85)**, 379–388.

Schrott, L., 1994. *Die Solarstrahlung als steuernder Faktor im Geosystem der Subtropischen semiariden Hochanden (Agua Negra, San Juan, Argentina)*. Ph.D. thesis, Universität Heidelberg.

Schwikowski, I., Barbante, C., Doering, T., Gaeggeler, H. W., Boutron, C., Schotterer, U., Tobler, L., Van De Velde, K. V., Ferrari, C., Cozzi, G., Rosman, K., and Cescon, P., 2004. Post-17th-century changes of European lead emissions recorded in high-altitude alpine snow and ice. *Environmental Science & Technology*, **38(4)**, 957–964.

Schwikowski, M., Doscher, A., Gaggeler, H. W., and Schotterer, U., 1999. Anthropogenic versus natural sources of atmospheric sulphate from an Alpine ice core. *Tellus Series B-Chemical and Physical Meteorology*, **51(5)**, 938–951.

Schwikowski, M., Rufibach, B., Schwerzmann, A., Stampfli, D., Barbante, C., Planchon, F., P., G., and Boutron, C., 2003. Two new ice cores from Colle Gnifetti, Swiss/Italian Alps. Technical report, Paul Scherer Institut, Laboratory for Radiochemistry and Environmental Chemistry, Annual Report 2003.

Siegenthaler, U., Schotterer, U., and Oeschger, H., 1983. Sauerstoff-18 und Tritium als natürliche Tracer für Grundwasser. *Gas-Wasser-Abwasser*, **63**, 477–483.

Siegenthaler, U., Stocker, T. F., Monnin, E., Luthi, D., Schwander, J., Stauffer, B., Raynaud, D., Barnola, J. M., Fischer, H., Masson-Delmotte, V., and Jouzel, J., 2005. Stable carbon cycle-climate relationship during the late Pleistocene. *Science*, **310(5752)**, 1313–1317.

Spahni, R., Chappellaz, J., Stocker, T. F., Loulergue, L., Hausammann, G., Kawamura, K., Fluckiger, J., Schwander, J., Raynaud, D., Masson-Delmotte, V., and Jouzel, J., 2005. Atmospheric methane and nitrous oxide of the late Pleistocene from Antarctic ice cores. *Science*, **310(5752)**, 1317–1321.

Stichler, W., Schotterer, U., Fröhlich, K., Ginot, P., Kull, C., Gäggeler, H., and B., P., 2001. Influence of sublimation on stable isotope records recovered from high-altitude glaciers in the tropical Andes. *Journal of Geophysical Research*, **106(D19)**, 22613–22620.

Suter, S., 2002. *Cold firn and ice in the Monte Rosa and Mont Blanc areas: spatial occurence, surface energy balance and climatic evidence*. Ph.D. thesis, VAW, ETH Zürich.

Suter, S., Laternser, M., Haeberli, W., Frauenfelder, R., and Hoelzle, M., 2001. Cold firn and ice of high-altitude glaciers in the Alps: measurements and distribution modelling. *Journal of Glaciology*, **47(156)**, 85–96.

Thompson, L., Davis, M., Mosley-Thompson, E., Sowers, T., Henderson, K., Zagorodnov, V., Lin, P., Mikhalenko, V., Campen, R., Bolzan, J., Cole-Dai, J., and Francou, B., 1998. A 25,000-year tropical climate history from Bolivian ice cores. *Science*, **282(5395)**, 1858–1864.

Thompson, L., Mosley-Thompson, E., Davis, M., LIN, P., Henderson, K., Coledai, J., Bolzan, J., and LIU, K., 1995. Late Glacial Stage and Holocene tropical ice core records from Huascarán, Peru. *Science*, **269(5220)**, 46–50.

Thompson, L., Yao, T., Davis, M., Henderson, K., MosleyThompson, E., Lin, P., Beer, J., Synal, H., ColeDai, J., and Bolzan, J., 1997. Tropical climate instability: The last glacial cycle from a Qinghai-Tibetan ice core. *Science*, **276(5320)**, 1821–1825.

Thompson, L. G., 1984. El Niño-Southern Oscillation Events Recorded in the Stratigraphy of the Tropical Quelcaya Ice cap, Peru. *Science*, **226**, 50–53.

Thompson, L. G., Mosley Thompson, E., Bolzan, J. F., and Koci, B. R., 1985. A 1500-Year Record of Tropical Precipitation in Ice Cores from Quelccaya Ice Cap, Peru. *Science*, **229**, 971–973.

Thompson, L. G., Mosley Thompson, E., Davis, M. E., Bolzan, J. F., Dai, J., and Klein, L., 1990. Glacial stage ice-core records from the subtropical Dunde ice cap, China. *Annals of Glaciology*, **14**, 288–297.

Veit, H., 2000. Klima- und Landschaftswandel in der Atacama. *Geographische Rundschau*, **52(9)**, 4–9.

Villa, S., Vighi, M., Maggi, V., Finizio, A., and Bolzacchini, E., 2003. Historical trends of organochlorine pesticides in an Alpine glacier. *Journal of Atmospheric Chemistry*, **46(3)**, 295–311.

Vose, R., Schmoyer, R., Steurer, P., Peterson, T., Heim, R., Karl, T., and Eischeid, J., 1992. *The Global Historical Climatology Network: long-term monthly temperature, precipitation, sea level pressure, and station pressure data. ORNL/CDIAC-53, NDP-041.* Carbon Dioxide Information Analysis Center, Oak Ridge National Laboratory, Oak Ridge, Tennessee.

Vuille, M., Bradley, R., Healy, R., Werner, M., Hardy, D., Thompson, L., and Keimig, F., 2003a. Modeling $\delta^{18}O$ in precipitation over the tropical Americas: 2. Simulation of the stable isotope signal in Andean ice cores. *Journal of Geophysical Research - Atmospheres*, **108(D6)**.

Vuille, M., Bradley, R., Werner, M., Healy, R., and Keiming, F., 2003b. Modeling $\delta^{18}O$ in precipitation over the tropical Americas: 1. Interannual variability and climatic controls. *Journal of Geophysical Research - Atmospheres*, **108(D6)**, 4174.

Wagenbach, D., 1992. Results from the Colle Gnifetti ice-core program, Report of the ESF/EPC Workshop (Greenhouse Gases, Isotopes and Trace Elements in Glaciers as Climatic Evidence of the Holocene), Zürich 27-28 October 1992. Technical report, Versuchsanstalt für Wasserbau, Hydrologie und Glaziologie, ETH Zürich.

Wagenbach, D. and Geis, K., 1989. The mineral dust record in a high altitude Alpine glacier (Colle Gnifetti, Swiss Alpsa). In M. Leinen and M. Sarnthein, editors, *Paleoclimatology and Paleometeorology: Modern and Past Patterns of Global Atmospheric Transport*, pages 543–564. Kluwer Academic Publishers, Dortrecht, Netherlands.

Wolff, E. W., Fischer, H., Fundel, F., Ruth, U., Littot, G. C., Mulvaney, R., Röthlisberger, R., De Angelis, M., Boutron, C., Hansson, M., Jonsell, U., Hutterli, M., Lambert, F., Kaufmann, P., Stauffer, B., Stocker, T. F., Steffensen, J. P., Bigler, M., Siggard-Andersen, M. L., Udisti, R., Becagli,

S., Castellano, E., Severi, M., Wagenbach, D., Barbante, C., Gabrielli, P., and Gaspari, V., 2006. Southern Ocean sea-ice extent, productivity and iron flux over the past eight glacial cycles. *Nature*, **440(7083)**, 491–496.

Zachos, J., Pagani, M., Sloan, L., Thomas, E., and Billups, K., 2001. Trends, rhythms, and aberrations in global climate 65 Ma to present. *Science*, **292(5517)**, 686–693.

Zachos, J. C., Lohmann, K. C., Walker, J. C. G., and Wise, S. W., 1993. Abrupt Climate Change and Transient Climates During the Paleogene - a Marine Perspective. *Journal of Geology*, **101(2)**, 191–213.

Die VDM Verlagsservicegesellschaft sucht für wissenschaftliche Verlage abgeschlossene und herausragende

Dissertationen, Habilitationen, Diplomarbeiten, Master Theses, Magisterarbeiten usw.

für die kostenlose Publikation als Fachbuch.

Sie verfügen über eine Arbeit, die hohen inhaltlichen und formalen Ansprüchen genügt, und haben Interesse an einer honorarvergüteten Publikation?

Dann senden Sie bitte erste Informationen über sich und Ihre Arbeit per Email an *info@vdm-vsg.de*.

Sie erhalten kurzfristig unser Feedback!

VDM Verlagsservicegesellschaft mbH
Dudweiler Landstr. 99
D - 66123 Saarbrücken
www.vdm-vsg.de

Telefon +49 681 3720 174
Fax   +49 681 3720 1749

Die VDM Verlagsservicegesellschaft mbH vertritt

Printed by Books on Demand GmbH, Norderstedt / Germany